Ecological Complexity and Agroecology

This text reflects the immense current growth in interest in agroecology and changing approaches to it. While it is acknowledged that the science of ecology should be the basis of agroecological planning, many analysts have out-of-date ideas about contemporary ecology. Ecology has come a long way since the old days of "the balance of nature" and other romantic notions of how ecological systems function. In this context, the new science of complexity has become extremely important in the modern science of ecology. The problem is that it tends to be too mathematical and technical and thus off-putting for the average student of agroecology, especially those new to the subject. Therefore this book seeks to present ideas about ecological complexity with a minimum of formal mathematics.

The book's organization consists of an introductory chapter, and a second chapter providing some of the background to basic ecological topics as they are relevant to agroecosystems (e.g., soil biology and pest control). The core of the book consists of seven chapters on key intersecting themes of ecological complexity, including issues such as spatial patterns, network theory and tipping points, illustrated by examples from agroecology and agricultural systems from around the world.

John Vandermeer is Asa Gray Distinguished University Professor of Ecology and Evolutionary Biology, University of Michigan, USA.

Ivette Perfecto is George W. Pack Professor of Natural Resources and Environment, University of Michigan, USA.

Ecological Complexity and Agroecology

John Vandermeer and Ivette Perfecto

LONDON AND NEW YORK

First published 2018
by Routledge
2 Park Square, Milton Park, Abingdon, Oxon OX14 4RN

and by Routledge
711 Third Avenue, New York, NY 10017

Routledge is an imprint of the Taylor & Francis Group, an informa business

British Library Cataloguing-in-Publication Data
A catalogue record for this book is available from the British Library

Library of Congress Cataloging-in-Publication Data
Names: Vandermeer, John H., author. | Perfecto, Ivette, author.
Title: Ecological complexity and agroecology / by John Vandermeer and
    Ivette Perfecto.
Description: Abingdon, Oxon ; New York, NY : Routledge, 2017. |
    Includes bibliographical references and index.
Identifiers: LCCN 2017020043 | ISBN 9781138231962 (hardback) |
    ISBN 9781138231979 (pbk.) | ISBN 9781315313696 (ebook)
Subjects: LCSH: Agricultural ecology.
Classification: LCC S589.7 .V359 2017 | DDC 577.4—dc23
LC record available at https://lccn.loc.gov/2017020043

ISBN: 978-1-138-23196-2 (hbk)
ISBN: 978-1-138-23197-9 (pbk)
ISBN: 978-1-315-31369-6 (ebk)

Typeset in Bembo
by Apex CoVantage, LLC
Printed and bound by CPI Group (UK) Ltd, Croydon, CR0 4YY

Dedicated to the memory of Richard Levins

Contents

List of figures   x
Preface   xviii

1 Introduction   1

The idea of ecological complexity 2
Early revolutionaries in the new agriculture 7
To "understand": nine reasons for generalization 12
A note on theory and intuition 14
Ecological complexity and the organization of this book 15

2 Basic ecological concepts   19

How plants get energy 19
How plants get nutrients from soil 24
Transformation of nutrients in the soil 27
Plants, soil and water 32
The ecological niche: a historical backbone 34
Population dynamics 39
Equilibrium, resilience, persistence 42
Basic trophic dynamics 47
Feedback in dynamic systems 50

3 The Turing mechanism and geometric pattern   60

Dynamic consequences of background exogenous pattern 64
Generation of endogenous pattern 69
Criticality and power functions 75
Percolation points and power functions 81
The skeleton of the fundamental niche, the flesh of the realized niche 84
Spatial pattern and the agricultural connection 86

4 Chaos   89

Introduction 89
Intuition of the importance of chaos in a simple system 93
Farmer math and the intuition of chaos 95
Chaotic attractors, transients and Cantor sets 99
Ecological chaos in the real world? 103
The generalized structure of chaotic attractors 106
From cooked carrots to chaos to attractor reconstruction 110
Conclusions 113

5 Stochasticity   115

Introduction: deterministic versus stochastic 115
The importance of stochasticity 116
Basic population processes with stochasticity 119
Predator/prey systems and stochasticity 125
The interrelationship between chaos and stochasticity 127
The chaos/stochasticity tapestry 130
Final comments 134

6 Coupled oscillators   135

An odd kind of sympathy 135
Consumer/resource oscillators and weak coupling: a basic pattern 136
Oscillatory structure in chaos: the teacup 139
Confronting Gause's principle with oscillations 140
Limiting similarity and species packing with oscillators 143
Decomposition as an oscillatory process 149
Seasonality and the Moran effect 152

7 Multidimensionality   157

Three-dimensional systems 158
Multidimensional systems 162
The promise of food webs 166
Qualitative structure of directed graphs – loop analysis 172
Concluding remarks 176

8 Trait-mediated indirect interactions   178

Density-mediated indirect interactions 179
Trait-mediated indirect interactions in principle: a basic nonlinearity 181
Trait-mediated effects in the real world 187

Consequences of trait-mediated indirect interactions 191
Hypernetworks 191
Multiple TMIIs 196
Summary 198

9 Critical transitions   200

Inevitability of surprise 200
Loss of biodiversity: expected or surprise? 201
Hysteresis on a global scale 205
The surprise of the coffee rust disease 208
Agricultural syndromes as hysteretic phenomena with tipping points 214
Basin boundary collisions: chaos and catastrophe 218
Summary 219

10 The "scientific" basis of agroecology   221

Complexity science and ecology 222
The four pillars of agroecology 224
The "whole" of agroecology 230
Solving farmers' problems in the age of ecological complexity 231
The importance of thought-intensive technology 233

References   235
Index   247

Figures

1.1 David Lavigne's partial food web with codfish as
the central player   3

1.2 Examples of complex systems   5

1.3 Conceptualizing ecological complexity as the intersection of
the seven overlapping topics thought to be important in
the field of agroecology   17

2.1 Light response curve in photosynthesis   21

2.2 Light response curves for C3 and C4 plants   23

2.3 The basic difference in light response curves between sun
plants and shade plants, and the the effect of photoinhibition   23

2.4 Basic process of cation exchange in plant nutrition,
illustrated with potassium dynamics; color gradient bubble
surrounding the root hair represents the concentration
gradient of potassium   26

2.5 Representations of the overall process of nitrogen dynamics   30

2.6 Simplified summary of the nitrogen cycle   31

2.7 Piston/cylinder metaphor of soil water dynamics   33

2.8 Cartoon representation of the state of soil water   33

2.9 Illustration of Hutchinson's framework for the ecological niche   36

2.10 Diagrammatic representation of the transformation of
fundamental to realized niche through the process
of competition   37

2.11 Illustration of the pattern of exponential growth of two
distinct populations   41

2.12 Illustration of logistic population changes   42

2.13 Stylized diagram of crop yield as a function of an overstory
of legume shade trees   44

2.14 Illustrations of the key concepts of resilience and persistence
with a physical model of ball on a surface   45

2.15 Representation of the basic oscillatory process as
"response" surfaces   46

2.16 Illustration of the time series resulting from the three
main types of equilibrium situations   47

2.17 General form of the predator/prey relationship
      (the fundamental trophic interaction)   49
2.18 Exploring oscillatory dynamics with isoclines   50
2.19 Transformation of the prey isocline with the addition of
      prey intraspecific competition   51
2.20 Adding intraspecific competition to the basic oscillatory
      pattern of trophic dynamics   52
2.21 Effects of relaxing the assumption of simple density dependence   54
2.22 Basic population dynamics of two species   55
2.23 Extension of Figure 2.22c, in which the expected response of
      (a) aphid 1 and (b) aphid 2 are shown independently, but
      with all densities of the competing species illustrated
      (with the horizontal or vertical vectors)   56
2.24 The classic four outcomes of interspecific competition   57
2.25 Response surface of a saddle point   57
2.26 The logic of alternative possible equilibrium points in a
      consumer/resource (predator/prey) situation   58
3.1   Vegetation frequently occurs in a non-random pattern
      of clusters   62
3.2   Clustered distribution of organisms   63
3.3   Landscapes from Brazil's Atlantic forest region   65
3.4   Huffaker's (1958) classic experiment   67
3.5   Qualitative representation of a predator/prey system in
      distributed space   68
3.6   Distribution of witch hazel (*Hamamelis virginiana*) stems in
      the E. S. George Reserve (University of Michigan) in
      a 21-hectare plot   69
3.7   a. A petri dish with a solution (white) into which a drop of
      activator chemical is introduced in the center, reacting with
      the solution to create a product (dark gray circle in the center),
      and begins diffusing across the petri dish (light gray area).
      b. The product (central black circle) reaches a critical
      concentration such that a repressor chemical is produced
      (stippled area), which begins diffusing (the original activator
      chemical continues to diffuse – shading indicates recent
      versus old diffusion area). c. The repressor chemical cancels
      the activator creating a zone of original solution (white),
      but continues diffusing, although its concentration gradually
      diminishes as it diffuses. d. The final result is the formation
      of a spatial pattern.   71
3.8   Typical computer-generated Turing patterns   71
3.9   Spatial distribution of the ant, *Azteca sericeasur*, in a 45-hectare
      plot within a coffee farm in Chiapas, Mexico   73
3.10 A proposed mechanism for generating vegetation patterns in
      arid and semi-arid environments   74

3.11   Field observations of the distribution of *Azteca sericeasur* nests
        in 2004 (a) and 2005 (b) and results of computer simulations
        with a Turing-based spatial model for the same data (c and d)   75
3.12   Example distributions of ant nests   76
3.13   Example of a self-organized critical system, the sand pile   77
3.14   The relationship between frequency of clusters as a function
        of cluster size, often described with the equation $y = ax^b$,
        where y is the frequency and x is the cluster size   77
3.15   Power function fits for the distribution of ant nests in
        Figure 3.12   78
3.16   Power function approximation to natural pasture vegetation
        in Spain   79
3.17   Diagram of the forest fire model, where shaded area indicates
        the spread of the forest fire   80
3.18   Natural log of the population density as a function of
        the density dependent local mortality rate (red curve)   83
3.19   The underlying basis for the pattern of Figure 3.2b (that is,
        imagine the points are samples of fish in the Great Lakes of
        North America)   85
4.1   Illustration of the principle of "sensitive dependence on initial
        conditions"   90
4.2   Illustration of population dynamics of two bacterial
        populations in spring (a and b) and summer (c and d)   91
4.3   The modeling of a convection roll   93
4.4   Farmer's math regarding pests   97
4.5   Projection of the population over a 15-year time period   98
4.6   The model of Figure 4.5 extended over a 100-year
        time period   98
4.7   Chaotic (strange) attractor of a population on two islands,
        with migration between them (on each island, the
        population sizes are scaled to be between 0 and 1.0,
        the fraction of the maximum population)   100
4.8   The same two-island attractor of Figure 4.7   102
4.9   Construction of a Cantor set, illustrating how, in the limit,
        there will be an infinite number of points (line segments),
        and 100% of the original area will be without points (empty)   103
4.10   Three time series illustrating chaos, randomness and an actual
        1,000-year time series of locust populations   104
4.11   Results of interviews with tomato farmers in Costa Rica
        where they were asked how much tomatoes they would
        plant in relation to current market prices   105
4.12   Relative prices of piglets in Japan from 1967 to 1992   105
4.13   A tornado touches ground near Anadarko, Oklahoma   106
4.14   Theoretical structure of an escaped transgene conferring
        resistance to attack from caterpillar pests through the Bt toxin   108

4.15   Chaotic attractor in N, N_m and P space, where N is the density
       of the non-genetically modified plant, N_m of the genetically
       modified plant and P of the pest   108
4.16   The Rossler attractor (the three dimensions are
       completely arbitrary)   109
4.17   The basic modification of the vector field to create a
       chaotic attractor   109
4.18   Essentials of "attractor reconstruction" (using the formality
       of Taken's theorem)   112
5.1   The coffee berry borer attacking coffee fruits, and the *Azteca*
       ant responding by removing the berry borer from the fruit   117
5.2   Illustration of the difference between variability and
       stochasticity, with the attack of the coffee berry borer
       occurring only when there is ant-free space, and only when
       it lasts for a continuous 60-minute period of time (each tick
       in time is 5 minutes)   118
5.3   Ten populations, with small growth rates, that are adjusted
       stochastically every year for a period of 100 years   120
5.4   Expected distribution of 50 distinct populations growing
       with an identical average growth rate, but with two distinct
       levels of stochastic forcing   122
5.5   The effect of adding stochasticity to a population that is
       characterized by a strong Allee effect   123
5.6   Results of adding stochastic forces to the simple model of
       density-dependent population growth   124
5.7   Illustration of the generation of stochastic limit cycle-like
       behavior in a predator/prey system   126
5.8   Results of a population growth model with a continuous
       variable (for number of pests per plant)   128
5.9   Results of a population growth model rounding the
       continuous variable (pest population size) to its nearest integer 129
5.10   Results of a population growth model adding stochasticity
       and then rounding the continuous variable (pest population size)
       to its nearest integer   129
5.11   Results of a population growth model adding a stochastic force
       and rounding to the closest integer, when the model is initially
       adjusted to predict precisely a 3-year cycle   130
5.12   *Tribolium castaneum* and an example of its environment   131
5.13   Diagrammatic illustration of the expected behavior of a
       population of beetles in three dimensions (A = adult densities;
       P = pupal densities; L = larval densities)   131
5.14   Larval population density of *Tribolium castaneum* over time,
       illustrating the two major modes of population behavior,
       early on approaching the unstable point and lingering there
       for a long period of time, then moving into the expected

two-level oscillations (not exactly the same two points, but one
high, followed by one low, followed by one high, and so forth)   132

5.15   Three dimensional plots illustrating the time series for
A = adults, P = pupae, and L = larvae of *Tribolium castaneum*
in the Jillson experiment (as shown in Figure 5.13)   133

5.16   Results of adding a stochastic factor to a basic
population model   133

6.1   Mosaic on front of building at Leidsestraat 88, Amsterdam   136

6.2   The variety of trophic oscillators common in nature   137

6.3   Simulations of weakly coupled predator/prey oscillators, with
the y-axis plotting the population density of consumer pairs
(bold trajectory is consumer 1, light trajectory is consumer 2)   138

6.4   The elementary coupling of consumer-resource systems
and the qualitative outcomes   138

6.5   Cartoon representation of a three trophic system (trophic
level 1 = aphid herbivore; trophic level 2 = lady beetle
predator of aphid; trophic level 3 = lizard predator of
lady beetle), as two coupled oscillators   139

6.6   The tri-trophic model of Hastings and Powell,
a chaotic attractor   140

6.7   Common, almost universal, arrangement of an organism
simultaneously consumed by a predator and a disease   141

6.8   Results of simulations of the simultaneous action of a
predator and a disease on a pest (S = susceptible pest,
I = infected pest, P = predator)   143

6.9   Elementary form of the Levins and MacArthur limiting
similarity argument   145

6.10   The two ways in which a third species can enter the
ecosystem of two independent oscillators such that
they are coupled through the third species   146

6.11   Control of an invasive species through an alternation of
in-phase and anti-phase coordination of the two consumers   147

6.12   Population density of all three consumers (R_1, R_2, R_3) in
the situation of resource competition   149

6.13   Bacterial density along the root of a wheat plant
(as a surrogate for time)   150

6.14   Generalized soil nitrogen dynamics and the production of
above-ground biomass of a species of plant   151

6.15   a. Relation between plant biomass and seed weight of wild
rice (the more biomass, the more organic matter to fertilize
the soil and feed the plants that then produce more seeds).
b. Identifying three clusters of points. c. Inferred population
dynamics of the wild rice population based on the three
clusters of points (qualitatively converting mean seed weight
per plant into total plant biomass of the next year, the ovals

indicate rough areas within which the population will
reside each time period), illustrating the possible oscillatory
dynamics that might arise from this relationship. A yearly cycle
suggested by the four bold arrows; a 5-year cycle indicated
by the 10 dashed arrows.   152

6.16 Physical model of the idea of a "forced" oscillator   153
6.17 Population responses to seasonal forcing   155
7.1   Illustration of the impossibility of trajectories crossing one
      another in two-dimensional space   158
7.2   Illustration of how a three-dimensional system permits
      trajectories to appear to cross when viewed in two dimensions,
      even if the third dimension is very small   159
7.3   Two typical weeds in many tropical agroecosystems   161
7.4   An example of the *dehesa* system in Bollullos Par del Condado,
      Huelva, southern Spain   162
7.5   Exemplary weedy plant species included in the community
      matrix in Figure 7.6   164
7.6   a. Estimated competition coefficients placed in a "community
      matrix." Relative rates of competition calculated from data
      in Goldberg and Landa, 1991. (Note that a negative value
      represents a facilitative effect.) b. Competitive outcomes
      matrix from (a).   164
7.7   a. The competitive outcomes network of the data in
      Figure 7.6b, based on the competitive outcomes matrix
      (1 = competitively superior; 0 = competitively inferior);
      b. The resulting graph if Lp is eliminated; c. The resulting
      graph after Rc is eliminated due to competition. Note that
      the resulting 5-species network includes four 3-species
      intransitive loops (indicated with different colors, patterns
      and thickness of line).   166
7.8   Idealized farm with a central region of some crop and banana
      plants in the periphery   167
7.9   The transformation of a network from regular to random   169
7.10 Representation of the coupled oscillator system presented in
      Chapter 6 (Figure 6.10b) as a network graph   170
7.11 Artificial example of an agroecosystem network,
      with 11 nodes   171
7.12 Pattern of network structure as a function of number of nodes
      randomly removed for a robust network, a non-robust network,
      and the network of Figure 7.11   172
7.13 a. Directed graph showing the sign of all connections,
      positive (with arrowheads), and negative (with small circles)
      and neutral (with no indication). b. The same graph with
      each direction of interaction (positive or negative)
      indicated separately.   173

7.14 Repeat of Figure 7.13b, with a loop of length 2 labeled
with dashed connectors and a loop of length 7 with bold
connectors   174
7.15 Basic structures involved in classical loop analysis   175
8.1 The trophic structure deduced from birds/bats exclosure
experiments in a Mexican coffee farm, generalized   181
8.2 *The Bird Scarer*, by William Knight Keeling (1807–1886),
painted between 1827 and 1886   182
8.3 Schematic of the effect of a second crop (the darker
plant) on an herbivore of the primary crop
(the lighter plants)   184
8.4 Flow diagram illustrating the transfer of energy and
the effect of trait-mediation in a simple system   185
8.5 Diagrammatic representation of the ant-scale-beetle-phorid
trait-mediated trophic cascade   190
8.6 Various network representations of the *Azteca* system that
include the phorid (P), *Azteca* ant (A), scale insect (S),
and beetle (B)   192
8.7 Density of adults and larvae of the scale-eating beetle,
Azya orbigera, as a function of distance to nearest nest of
the ant *Azteca sericeasur*   193
8.8 Representation of the control of the scale insect pest in
the coffee agroecosystem   194
8.9 Illustration of the two classes of TMII in a generalized
trophic situation   195
8.10 TMIIs with alternative effects on the transfer of energy from
gazelle to cheetah   195
8.11 The framing of the higher-order effects using the
framework from Figures 8.4 and 8.5   197
8.12 Example of the general outcome of simulations
with α, β and γ connections   197
9.1 Biodiversity changes in response to agricultural intensification 202
9.2 The construction of tipping points and hysteresis   204
9.3 Examples of the qualitative relationship between ecological
regime and ecological conditions   206
9.4 Relationship between annual precipitation and percent tree
cover creates distinct ecological regimes   208
9.5 Basic life cycle of the coffee rust disease   210
9.6 Two extremes of coffee production systems   211
9.7 Illustration of the changes in expectation of the coffee rust
disease, assuming that the transmission of spores is the only
factor affecting the disease   213
9.8 The expected relationship between coffee rust disease and
the amount of shade cover in a coffee agroecosystem,
based on an underlying model that includes two spatial

scales of transmission, shade-induced transmission and
germination effects, and potential role of natural enemies   214
9.9   The farm response emerging from the conditions of production
framed as three distinct syndromes of production, that is, three
alternative stable states   217
9.10 Time series trajectory of chaotic attractors   219

Preface

Our experiences in writing this book are sandwiched between our colleague, comrade, and mentor, Richard (Dick) Levins, on the one hand, and the current rural social movements spanning the globe, especially the Zapatista movement of southern Mexico. Levins certainly set the stage for both of us, independently of one another, and the Zapatistas' recent articulation with science has reinforced our notion that science should be aimed at the needs of people, not profit. It is worth a couple of short notes introducing these influences on the two of us.

Ivette was a university student, studying biology in Puerto Rico during the 1970s. She was actively involved in the student political movements supporting independence for Puerto Rico and had heard stories of the *gringo* who was a professor at the University of Puerto Rico (UPR) and was a strong supporter of the student movement for independence of Puerto Rico. For his political view, Richard Levins, the *gringo* professor did not get tenure at the UPR and went back to the United States where he got a faculty position first at the University of Chicago and then at Harvard. She never met that professor but developed an admiration for his solidarity with the Puerto Rican struggle. Several years later she went to graduate school at the University of Michigan to study ecology. There, she was introduced to ecological theory and the seminal contributions of Richard Levins to modern ecology. She met Levins when he went to give a seminar in Ann Arbor, and in conversations with him she realized that he was the same *gringo* professor that did not get tenure at UPR for his political position in support of Puerto Rican independence. Realizing this identity was the beginning of her journey into the complex interactions between the political and the ecological, and getting to know Levins personally later, was part of that journey. As a graduate student at Michigan she joined Science for the People and the New World Agriculture Group (now New World Agriculture and Ecology Group; NWAEG), both of which were groups of scientists interested in social justice and applying their science for the benefit of the people. Levins and Lewontin's book *The Dialectical Biologist* has a strong influence on her academic and political development moving forward.

John's interactions with Richard Levins started earlier. He was a graduate student interested in mathematical ecology, focused at the time on laboratory experiments with microbes (it seemed, at the time, that ciliates were the biggest things that might look like differential equations). A trip to Costa Rica to take a course in tropical biology in 1968 got him out of the laboratory for good, and a visit of Levins to Michigan at that time so impressed fellow graduate students (John was in Costa Rica at the time) that they urged John to apply to do a post-doc with him. One postdoctoral year with Richard Levins is a rather full intellectual experience, to say the least. Intending to develop his skills in mathematical ecology, the realities of the Vietnam War were front and center at the University of Chicago at the time and discussions of political issues could not be avoided. A field course with Levins on the islands surrounding Puerto Rico was especially important, where interminable conversations about theoretical ecology, political revolution, art, music, activism and other subjects became theses and antitheses, interpenetrating opposites, in the struggle to make academic life meaningful.

What was important to the both of us was the complete commitment to an intellectual engagement of all aspects of life. Levins was not one to partition either his personal or academic life. Organizing a protest march received the same amount of intellectual and emotional attention as solving the equations of fitness sets. It is this wholeness of life that is difficult to explain, but for both of us was evident in our experiences with Dick over the next 45 years before his death. No doubt his general life philosophy is reflected in this work as much as his evident intellectual mark.

The contemporary world is filled with important signposts marking what we hope will be part of the future. Of particular importance presently, the indigenous peasant movement known as the Zapatistas looms large for us. As we were in the final stages of writing this book we were invited to participate in an event organized by the Zapatistas called *ConCiencias* (in Spanish it has the double meaning "With Science" and "Consciousness"). This difficult-to-categorize movement is clearly anti-capitalist, but above all is engaged in the construction of a new, more just, world. And in that world they see science as one of the fundamental pillars of the new society. Although the Zapatistas have a critical approach to Western science, and in particular science that is used explicitly in the service of capitalism, they don't throw the baby with the bathwater. They recognize the power of science and its importance as a tool for intellectual self-defense, for resistance, and for the construction of that new world that they envision.

For us, participating in this gathering was a bit ironic since it was at the same time that the US political system veered dramatically to the right with, astonishingly, the election of a president who questioned the utility of science at all. An "ignorant" peasant political movement in the highlands of Chiapas seemed to us to more enlightened than many of the citizens of the most powerful country in the world.

The *ConCiencias* meeting was symbolic. It represented a position that we have long held theoretically, that science should serve the needs of humanity, and that one of those needs was to develop a sociopolitical system that indeed promotes that service. It is a position and a praxis that is opposed to the model of neoliberal capitalism, pioneered by Ronald Reagan and Margaret Thatcher and their subsequent enthusiastic supporters. Our personal connection to this position regarding science is primarily through agriculture. Since our training is in ecology and our passion is with the natural world, the burgeoning movement of agroecology provides us with a clear pathway to our present positions.

With this political background we must also acknowledge the importance of the natural world to our point of view. We both have been "nature lovers" for most of our lives and came to much of our active research through the lens of conservation. Both scientific and popular literature send the same depressing signal about the continuing loss of the world's biodiversity. Experiencing this mainly through our work in tropical America, we could hardly ignore the complex relationships between so-called preserved areas and the agriculture surrounding them. The continued changes in agriculture, promoted mainly by the underlying assumption that land should be devoted to the profit motive, was evident in this part of the world that harbors most of the world's biodiversity. The savage assault on the natural world by neoliberal capitalism was hard to ignore. Yet the importance of the agricultural system to all of this was also hard to ignore, as we argued in our two previous books, *Nature's Matrix* and *Coffee Agroecology*.

Adding to this mix of influences, the explosive growth over the past 30 years in what has come to be called "complexity science" has been an important element in our intellectual growth. While the development of this field has been most evident in engineering and the physical sciences, its potential importance to the field of ecology is recognized by most contemporary ecologists. It is a rather math-intensive field. Nevertheless, as attested to by many popular articles on the subject, the results of complexity theory can be stated in a qualitative fashion. Our experience in teaching the subject matter suggests that the true significance of the insights provided by complexity science can be transmitted without the math. Especially with regard to the particular themes we think are relevant to the growing field of agroecology (i.e., the seven themes included in this book), the potential significance of complexity science can be presented with virtually no mathematics involved. The subjects are sometimes difficult to understand, but that is the nature of complex systems, with or without mathematics.

If the intersection of sociopolitical and conservation concerns motivated our first book (*Nature's Matrix*), and our detailed study of one of the important agro-ecosystems that illustrates that intersection motivated our second book (*Coffee Agroecology*), the importance of ecological complexity for the future of agriculture motives this third book, of what we like to think of as a trilogy. Adopting the idea, promoted by many others throughout the world, that the science of ecology should be the science that underlays agriculture, we asked, what is it about ecology that matters? Certain themes in agriculture are obvious – biological control of pests, supplying nutrients to crops, the emergence of pathogens, and so forth. In other words, the basic themes of ecology are obviously relevant to agriculture – predator–prey cycles, plant competition, nutrient cycling, trophic structures, and others. But the scientific field of ecology did not stop developing in 1960. Over the last few decades, ecology as an academic discipline has been influenced, perhaps even transformed, by the science of complexity. That transformation has occurred largely within the hallowed halls of the academy, effectively hidden from view with regard to potential applications to the field of agroecology. Unlike epidemiology, which has embraced the science of complexity with a passion, agroecology has not taken full advantage of modern developments in ecology. Speaking to this potential, we propose that agroecosystems are complex systems and as such they can be best understood with the tools that the science of complexity provides us. This book is our humble, and certainly incomplete, attempt to bring complexity science to bear on agroecology.

We started this preface describing the two main inspirations for this book, our teacher, mentor, friend and comrade, Richard Levins, and the Zapatistas indigenous movement. However we can't finish this preface without also acknowledging the contributions of all of our students, postdocs and visiting scholars who throughout the years have challenged and inspired us, namely, Saul Alarcón, David Allen, David Andresen, Bolivar Aponte, Inge Armbrecht, Katia Aviles, Cathy Bach, Bob Barretto, Monique Barry, Brent Blair, Janine Bologna, Heather Briggs, Justine Burdine, Jahi Chappell, Patrick Christie, Sarah Cohen, Julie Cotton, Diane DeSteven, Kay Dewey, Tom Dietsch, Evandro do Nacimento Silva, Elizabeth Dorgay, Iracenir Dos Santos, Bill Durham, Naim Edwards, Kate Ennis, Bruce Ferguson, Kaleigh Fisher, Gordon Fitch, Luis Garcia-Barrios, Miguel Antonio Garcia, David Gisaru, Paul Glaum, Bob Glesener, David Gonthier, Katie Goodall, Iñigo Granzo de la Cerda, Dan Griffith, Mike Hansen, Zach Hijian-Forooshani, Chau Ho, Hsun-Yi Hsieh, Aaron Iverson, Eliot Jackson, Julie Jedlica, Shalene Jha, Esteli Jimenez, Lilie Kline, Sergio Knaebel, Stefanie Krantz, Ashley Larson, Jesse Lewis, Kevin Li, Heidi Liere, Brenda Lin, Andy MacDonald, Maria Antoinia Mallona, Linda Marin, Austin Martin, Alex Mas, Stacy Mates, Krista McGuire, Hugh McGuinness, Geoffrey Michael, Zack Miller, Jeremy Moghtader, Helda Morales, Jonno Morris, Kristen Nelson, Virginia Nickerson, Theresa Ong, Beatrice Otero, Ethel Peternelli, Stacy Philpott, Chris Picone, Jarvas Queiroz, Jane Ramfert, Les Real, Marg Reeves, Katie Richter, Iris Rivera, Sandra Rodriguez, Peter Rosset, Dana Roth, Javier Ruiz, Edmond Russel, Andrea Samulón, Katherine Savoie, Lauren Schmidt, Erika Schreder, Brian Schultz, Hailey Schurr, Carolina Simao, Jane Skillman, John Soluri, Eduardo Somarriba, Lorena Soto Pinto, Zu Tan, Casey Taylor, Shinsuke Uno, Jerry Urquhart, Mariana Valencia, Chatura Venata, Kimberly Williams-Guillén, Elizabeth Witt, Katherine Yih and Senay Yitbarek. Special thanks to our colleagues/comrades Aaron King, Pej Rohani, Mercedes Pascual, Miguel Altieri, Clara Nichols, Angus Wright, Deborah Goldberg, Catherine Badgley, Gerry Smith, Braulio Chilil, Gustavo Lopez and Walter Peters, whose influence on our thinking shows through clearly.

1   Introduction

A new enthusiasm grips agriculture today. Across the globe old and young farmers are moving toward rethinking the foundations, rejecting the old triumphalist story of "industrial agriculture" to embrace the idea that the science of ecology should be at the foundation of agriculture, much as it has been in more traditional forms of agriculture in the past and present.[1] It is an enthusiasm, transforming to a revolution, born from a careful and critical analysis of the flawed story of how the industrial revolution came to agriculture to provide the world with ever-increasing productivity based on sophisticated technology – generating a gigantic cornucopia of enormous quantities of food. The ugly underbelly of this behemoth has become too difficult to ignore, with cancer epidemics produced from its pesticides, ocean dead zones fueled by its fertilizers, and farmers representing generations of tradition expelled from their land and even their culture. Yet the modern industrial agricultural system seems incapable of facing up to the cul-de-sac it has created for itself, producing hunger and obesity simultaneously, degrading soils, creating dead zones in the oceans, and knowingly promoting technologies that worship short-term gain over long-term sustainability. It is thus not surprising that a revolution would emerge.

Resisting the narrative of the glorious victory over nature, this movement to divest from the industrial agricultural system has been waiting in the wings at least since the nineteenth century, patiently accumulating sustenance from experience and analysis.[2] And its location is not in the centers of political and economic power, but rather in regions where traditional knowledge (frequently connected to popular social movements) intersects with the natural world.[3] Emerging as practice at the level of the farm, to some extent refusing modern scientific accoutrements, this new revolution has one central theoretical tenet that runs throughout: the fundamental rules of natural systems should be guidelines for planning and management. Or, equivalently, the science of ecology should be foundational, the key observation that motivates this book. And what precisely is that science?

Ecology is not an old and venerated discipline. Even though Darwin himself wrote brilliantly about it, he did not even recognize it as a thing. Ernst Haeckel is credited with inventing the word,[4] but certain key concepts emerged only in the twentieth century, which makes it little more than a century old. Yet that century provided key observations, experiments and theories that today are regarded as much a scientific core as Newton's theory of gravity is for physics. Especially in the past three decades, modern ideas have been brought to the table and what had been moving in a purely "ivory tower" direction suddenly appears poised to merge with more traditional ideas. One of those modern ideas is "complexity," and an ecological system is increasingly recognized as a "complex system." Originally from the Greek *com*, meaning together, and *plex*, meaning woven, a complex system is characterized by its "interwovenness."

Although complexity (and complex systems) is generally regarded as *avant garde* in academia,[5] we find that it is not a foreign concept to most traditional farmers we have encountered. Indeed our experiences include poor farmers in Latin America explaining elements of how their farm works in ways that, with only a change in some of the particular words used, could be mistaken for the introductory paragraph of a textbook of complex systems. They speak of, for example, chaos in product pricing, although their lexicon does not admit the word chaos. They understand that predators and their prey interlock such that long-term oscillations may emerge, although they would not call those coupled oscillators. They know that fixing the exact time for planting requires them to make environmental observations that have important unpredictable components, but we have never heard a farmer say "stochasticity." In the historical junction at which we seem to be poised, we argue that the time is ripe to take a burgeoning interdisciplinary movement within academia, complexity science, and combine it with various traditional forms of knowledge to construct a unique paradigm to undergird the burgeoning movement called agroecology.

In sum, the past quarter century has seen the field of ecology become enriched with ideas borrowed from complexity science, generating a new kind of ecology, increasingly referred to as "ecological complexity." Exploring the intersection of this new ecology with insights from traditional systems, in service of understanding the agroecosystem, is the fundamental subject of this book.

The idea of ecological complexity

It is not likely to be controversial to suggest that ecological systems can be very complicated. A useful example we have been using in teaching for years is the cod fishery in Newfoundland. In the early 1990s the fishery collapsed, leading to an energetic search for culpability. And some of the Canadian popular press hyped the idea that seals, known predators of codfish, were at fault, effectively taking a simplified view of the ecosystem as "seals eat codfish," nothing more to say. Canadian environmentalist David Lavigne looked more carefully at the system and concluded that the situation is a bit more complicated, to say the least. His graph, now widely cited when reporting on the amazing entanglement of ecological systems, is reproduced here in Figure 1.1.

Yet the largely correct observation that ecological systems are complicated is really quite different from the claim that they are complex, at least in the modern usage of those terms, especially as we use them in this book – complicated does

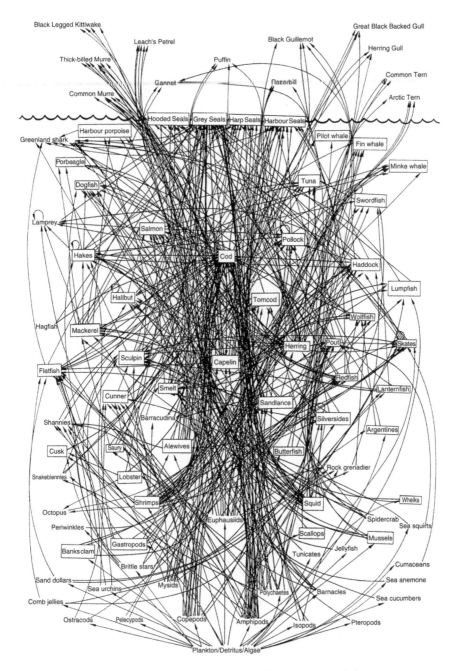

Figure 1.1 David Lavigne's partial food web with codfish as the central player

Compiled from a variety of sources by D. Huyck and reprinted from Lavigne (2003).

not equal complex. A complicated system is a system with many components; by understanding each component you can understand the whole. In contrast, as mentioned before, a complex system is characterized by interdependencies and it is the relationships among the parts that give rise to the whole. Understanding each component does not necessarily lead to an understanding of the whole. Complex systems have, in recent years, taken on a meaning that is more sophisticated and subtle than in the past. Examples of what current "complexity" scientists regard as complex systems range widely, with applications in disciplines ranging from physics and chemistry to sociology and economics, to neurobiology and behavior and, of course, ecology.[6] But the statement that a system is complex means more today than simply that it has many components (i.e., that it is complicated).

A general flavor of the idea can be appreciated with a key contradiction. It is certainly common sense to think that a large number of things interacting with one another will give rise to a pattern or a behavior that is itself quite complicated, and this seems to be generally true. It is likewise common sense to think that a very small number of things interacting with simple rules will give rise to a pattern or a behavior that is itself quite simple, and this too seems to be generally true. Imagining a 5-year-old child drawing at her desk evokes relatively simple images of behavior, while imagining a classroom of twenty 5-year-old children interacting with one another evokes relatively complicated images of behavior. Yet this sensible description of the world can be surprisingly contradicted by systems that are truly complex. To take a classic pair of ideas, when systems are truly complex it is sometimes the case that (1) apparently simple systems can exhibit very complicated behavior and (2) apparently complicated systems can exhibit very simple behavior. Frequently complex systems, as viewed through a contemporary lens, take on one or another of these forms, sometimes appearing to incorporate both at the same time. The seemingly straightforward correlation between complicated structure and complicated patterns or behavior is in contradiction with the actual potentialities of complex systems to generate unexpected and unusual patterns and behavior.

Consider, for example, the whirling and ever-changing patterns of flying starlings in their famous murmurations (a beautiful example can be seen, as of this writing, at: www.youtube.com/watch?v=eakKfY5aHmY), an image of which we reproduce in Figure 1.2a. Such spectacular moving patterns formed by these birds can be explained by nothing more than individual starlings simultaneously trying to locate themselves at the center of the group of individuals surrounding them and trying to stop anyone near them from getting too close. Thus, a spectacularly complicated pattern emerges from two very simple rules.

At the other extreme, ecological communities are normally very complicated (recall Figure 1.1), and spatially distributed in a non-random fashion. Multiple interactions occur among species; extinctions of species happen regularly; new species arrive from afar; predation, competition, mutualism and many other forces are operative. Yet that complexity can yield a remarkably simple pattern. A classic example is the structure of species diversity on islands (Figure 1.2b).

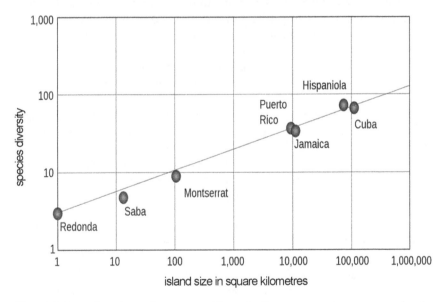

Figure 1.2 Examples of complex systems. Above: Starling murmurations, an example of a system with very simple rules producing complicated behavioral patterns. Below: The species diversity/island size relationship, caused by very complicated patterns of reptile and amphibian species interactions on and among islands in the Caribbean, yet forming a regular and predictable relationship between species diversity and island size.[7]

Multiple species interact with one another as competitors, predators, behavior alterers and more, with population-level processes such as extinction and migration imposed on top of such a complicated system, yet a very regular linear pattern emerges when the log of the number of species is plotted against the log of the size of island. A regular and simple pattern emerges from extremely complicated rules.

As is evident from the previous examples, a central feature of complex systems is that you cannot understand the system as a whole by breaking it down into smaller parts and studying those parts in isolation. Although any scientific endeavor must simplify a system in order to study it, the system itself cannot be simplified – it takes on its characteristics from the interactions of its components. We cannot describe it any better than Levins and Lewontin already described it in their classic *The Dialectical Biologist*:

> It is not that the whole is more than the sum of its parts. But that the parts acquire new properties. But as the parts acquire properties by being together, they impart to the whole new properties, which are reflected in changes in the parts, and so on. Parts and wholes evolve in consequence of their relationship, and the relationship itself evolves. These are the properties of things we call dialectical: that one thing cannot exist without the other, that one acquires its properties from its relation to the other, that the properties of both evolve as a consequence of their interpenetration.[8]

Twenty years later, political scientists Miller and Page place the same issues within the paradigm of the "new" science of complex systems:

> If parts are really independent from one another, then even when we aggregate them we should be able to predict and understand such "complicated" systems. As the parts begin to connect with one another and interact more, however, the scientific underpinnings of this approach begin to fail, and we move from the realm of complication to complexity, and reduction no longer gives us insight into construction.[9]

It is this vision of complexity that we hope to bring to the theory and practice of agroecology. Although remaining in the shadows of the industrial agricultural system that dominates most agricultural land in the world, it is striking, in some regions, how rapid is the growth of agroecological farms, even if not specifically called "agroecological," and despite the fact that the actual land surface they occupy remains small.[10] However, it is the spirit that these farmers bring with them that is the most striking. Millions of farming families are actually working the land, caring for the land, and in many cases taking on the mantra of being involved in something new, something perhaps revolutionary. This newness derives in part from a negation of the large-scale industrial monocultures, from a realization that alternative forms of production must, as soon as possible, replace the wasteful and environmentally poisonous system that emerged from the industrial agricultural model. These ideas, like all ideas, originate neither in a political nor historical vacuum. The intersection of ideas from multiple actors, acting across generations, is what gives this new movement its strength. We begin here by acknowledging a few of these actors so as to recognize at least some of the giants on whose shoulders we sit.

Early revolutionaries in the new agriculture

George Washington Carver is one of the most underappreciated figures in the history of US agriculture. He was a scientist, an educator and a social activist.[11] He started his life as a slave, yet climbed the ladder of academic success, securing a faculty position at the Tuskegee Institute in Alabama. He was dedicated to the education of the next generation of African American students and, most impor-tantly, established a system of bringing science to the people with his mobile demonstration laboratories (so-called Jessup wagons). It was only through him that poor black farmers were able to gain access to scientific information about agriculture. He set the stage for the transfer of scientific information, including the latest scientific discoveries, to the poorest of the poor in the United States.

While Carver brought science to the farmers, Gabrielle Matthaei and Albert Howard learned from the farmers and brought their deep multigenerational knowledge to scientists.[12] These two British botanists, married in 1905 and intimately involved in each other's research, were dispatched to Britain's vari-ous colonies to teach "the backward farmers of the colonies" about the newest advances in agriculture honed in the agricultural experiment stations of the British Empire. But when they got to India, they realized that the Indian farmers had agricultural techniques that were superior in many ways to the agricultural techniques developed in Britain's experiment stations. They decided to learn from, rather than teach to, the Indian farmers. Some of those agricultural prac-tices had been around for generations and had been refined by hundreds if not thousands of years of trial and error. The accumulated agricultural traditions of generations of Indian farmers had produced a wealth of knowledge about agri-culture that was simply unknown in the venerated British experiment stations. Matthaei and Howard set the stage for the transfer of traditional knowledge to the broader collection of scientific knowledge about agricultural production.

Based, in part, on influence of these two British botanists, Lady Eve Balfour converted her large farm to organic production, and began a series of long-term observations. Known as the Haughley Experiment, her observations are described in her book *The Living Soil*.[13] This was the first time that detailed observations were made on organic production on a whole farm level, and many of Balfour's observations have become conventional wisdom in today's organic agriculture movement. In 1946, inspired principally by Balfour's book, the Soil Association was formed in Great Britain, the first organization specifically devoted to the promotion of agricultural methods that would be self-sustaining, what we would call today organic or ecological agriculture. Its stated aim, at the time it was founded, was "to create a great body of biological knowledge of the life of the soil and to distribute that knowledge far and wide to the consumer as it accrues to the cultivator." The society still exists and is the main agency for the establishment of standards for organic produce in the United Kingdom.

Much earlier than Carver, Matthaei, Howard or Balfour, a group of peasant farmers in seventeenth-century Britain established an important agrarian tradition. Forming an organization originally referred to as True Levelers, but eventually reaching iconic status as the Diggers, they explicitly challenged the authority of monarchy, demanding that the Crown stop "defiling" mother earth and cede legitimate power to those who dig the earth, the farmers.[14] As an organization, they set the stage for many subsequent activist rural movements that challenge illegitimate political power over land and unfair land distribution. They developed a radical agrarian program based on the idea that "true freedom lies where a man (sic) receives his nourishment and preservation, and that is in the use of the earth." From their "Declaration of the poor oppressed people of England" they note, in part,

> [we] give notice to every one whom it concerns, . . . that some of you, that have been Lords of Manors, do cause the Trees and Woods that grow upon the Commons, . . . to be cut down and sold, for your own private use, thereby the Common Land, which your own mouths doe say belongs to the poor, is impoverished, and the poor oppressed people robbed of their Rights, while you give them cheating words, . . . Therefore we are resolved to be cheated no longer, nor be held under the slavish fear of you no longer, seeing the Earth was made for us, as well as for you. . . . . Therefore we require, and we resolve to take both Common Land, and Common Woods to be a livelihood for us, and look upon you as equal with us, not above us, knowing very well, that *England* the land of our Nativity, is to be a common Treasury of livelihood to all.

The Diggers may have been short lived, but their ideology lives on as a militant attachment to the land and the right to use it, and a rejection of illegitimate power, especially a power that cynically abuses and sometimes destroys the "Common Land," or in modern parlance, the natural world. Perhaps the Diggers were the first agroecologists.

Finally, the most recent of our influences is Rachel Carson, best known for her whistleblowing about the dangers of pesticides in her influential book *Silent Spring*. But Carson's contribution in the context of the present text preceded her most famous work. In *The Sea Around Us* and *Between Land and Sea* she anticipated the whole idea of ecological complexity,[15] in our view. Although one could argue that such anticipation had been manifest much earlier by such authors as Engels and Darwin, Carson's prose reflects, in our view, a sensibility to the sorts of complexities that go beyond "complicated," anticipating the modern idea of "complexity."

We won't idealize these men and women on whose shoulders we seek to perch. As all of us, they were people of their times, with views and ideas that may seem antiquated from today's perspective. However, they all challenged the conventional wisdom of their times and articulated a new vision for agriculture, and for that reason they represent the nascent structure of the alternative agricultural revolution, the agroecological revolution.

When we visit small-scale farms in both North and South, from Canada to Chile, from California to Kerala, there is a theme that recurs: "My farm is a part of a bigger nature and is not immune to the rules that govern that bigger nature." And those "rules that govern nature" are the core of the intellectual tradition called ecology. Thus the clarion call for "agroecology" suggests to us that the science of ecology is, or should be, the basis for planning a farm.

The Diggers set the stage for radical political action, reflected in the various peasant movements operative today, and Carver established the practice of bringing the latest scientific advances to the people, to the farmers who most needed to understand them. We strongly advocate in favor of the many "new ruralities"[16] that are being invented and pursued by small farmer organizations, especially those associated with the global peasant organization, La Via Campesina,[17] and our inspiration to bring the results from modern ecology to the table in support of those pursuits comes, at least in part, from Carver's example. But our detailed approach to the purely ecological side of the issue is most influenced first by Albert Howard and later by Rachel Carson.

Although the ecological principles that might be used to enlighten our understanding of agroecosystems, the subject of this book, are eclectic and somewhat complicated from a technical point of view, some of the principles upon which our understanding depends were established long ago, and probably multiple times. As an example we take the insights that Albert Howard clearly established in his book *An Agricultural Testament*, published in 1940, 10 years after his wife Gabrielle Matthaei passed on. It would be difficult to imagine that her influence on his research was not reflected in this important work, even though she is not formally an author. Two sets of imperatives emerge from this book as important for our purposes. First, we must understand where the knowledge of agroecosystems comes from in the first place, and second, we must understand how agroecosystems function when they are functioning well. Regarding the first imperative, Howard notes:

> In the study of soil fertility [*and agroecology in general*] the first step is to bring under review the various systems of agriculture which so far have been evolved. These fall into four main groups: the methods of Nature – the supreme farmer – as seen in the primeval forest, in the prairie, and in the ocean; the agriculture of nations which have passed away; the practices of the Orient [*and the Global South in general*], which have been almost unaffected by Western science; and the methods in vogue in regions like Europe and North America to which a large amount of scientific attention has been paid during the last hundred years."[18]

Note that Howard views the source of information broadly and, effectively, seeks the intersection of all sources of knowledge.

In addition to his insights on sources of information, he was also, in our view, one of the originators of what some more contemporary advocates have termed "natural systems" agriculture (farming with nature, ecologically based farming, etc.), effectively the principles that seem to apply to farms that are functioning well. He discussed eight general principles drawn from Indian farming practices:

1   Mother earth always raises mixed crops
2   Great pains are taken to preserve the soil
3   Vegetable and animal wastes are converted into humus
4   There is no waste
5   Processes of growth and decay balance one another
6   Maintain large reserves of fertility
7   The greatest care is taken to store the rainfall
8   Both plants and animals are left to protect themselves against disease.

We propose that these fundamental ideas remain true today, not only to characterize those farms that are functioning well (sustainably, productively, etc.), but as approximate guidelines for planning a farming operation. Our purpose in this book is to relate them to recent progress in the academic field of ecology, focusing specifically on the potential for increased understanding of how current agroecosystems operate, and as a guide for planning future agroecosystems. But we do not suggest that Howard's fundamental principles have somehow changed. They remain, we believe, solid statements of how agroecosystems function when they are functioning well. Recent advances in the science of ecology merely enrich the fundamental principles that the Indian farmers taught Matthaei and Howard.

Carson, noted mainly for her criticism of pesticides, significantly enriches our argument with her deep appreciation of indirect and nonlinear interactions, reflected in her beautiful prose describing the natural history of the sea and seashore. Her two books prior to *Silent Spring* (*The Sea Around Us* and *Between Land and Sea*) are filled with natural history anecdotes that feature what today could be called ecological complexity. While her development did not include the more technical issues we deal with in this text, her inspiration from nature led her to anticipate how complexity, not simply complications, would be important in our continual attempts to understand nature's nature. Her perspicacity provides us with a platform for generalization, retrospectively applying to her insights some generalizations that have emerged in the field of ecology over the past couple of decades.

Bringing materialism into ecology

The intersection of Carver, Matthaei, Howard, Balfour, Carson and the Diggers form the foundation of what we anticipate will be the evolving future of agroecology. Yet it is more than of passing interest to appreciate where some of the contemporary ecological framings come from and how they evolved, especially in light of some recent popular notions of what ecology constitutes. Predominant in what seems to us to be some elementary confusion is an almost romantic metaphor of nature as some sort of creature.

The history of ecology is replete with philosophical debates that sometime seem esoteric and unimportant. However, understanding these debates and where they originated (i.e., who proposed what and what were their political influences) can provide some insights about the actual ideas themselves. For example, in the 1920s and 1930s a major philosophical split occurred in ecology. When the American plant ecologist Frederic Clements proposed the idea of the "ecological community as an organism" in 1916, it was criticized as a throwback to Platonic Idealism. At around the same time Jan Christian Smut, a British philosopher, military man and, at that time, prime minister of the Union of South Africa, was developing the concept of "holism" that he described as "the tendency of nature to form wholes that are greater than the sum of the parts through creative evolution." Smut's ecological holism strongly influenced the South African grassland ecologist John Phillips, who incorporated humans as part of the natural world and argued that biotic communities were harmonious. These ideas fit well with Clement's notion of the ecological community as an organism and are considered part of the same idealist influences in ecology.[19] However, Smut's influence on Phillips also included the idea that human beings were naturally organized in racial hierarchies and led him to support the racist biocentrism of Smut, including the proposition that the races should be separated, an obvious preamble to apartheid. In opposition to the idealist philosophy of Clements, Smut and Phillips was the materialism of British Marxist scientists like Lancelot Hogben and Hyman Levy. In the famous "Nature of Life" debate during the 1929 meeting of the British Association for the Advancement of Science in South Africa, Hogben proffered his materialism as an alternative to Smut's holism and accused him of using holism to promote racist ideas. Interestingly, holism was later resurrected within a materialist framework by Marxist ecologist Richard Levins with the quasi-Hegelian "The truth is the whole." But, according to environmental historian John Bellamy Foster,[20] at the time the strongest opposition to the ecological holism of Smuts was articulated by the leading British plant ecologist Arthur Tansley. Tansley was a student of Ray Lancaster, a zoologist, evolutionary biologist and socialist, and was highly influenced by his mentor. In an article published in the journal *Ecology* in 1935, Tansley attacked the idea of the super-organism of Clements and the holism of Smut and Phillips and introduced the concept of the ecosystem, using a materialist conception of dynamic equilibrium of natural systems that was highly influenced by Levy. The ecosystem concept of Tansley included the biotic and abiotic environment and was meant to be understood through close empirical analysis rather than some pre-supposed intrinsic purpose, as the 'biotic communities' of Smut's holistic ecology (Bellamy Foster et al., 2010). The vision of ecological "communities" and "ecosystems" as dialectical complex systems in Levins and Lewontin's *The Dialectical Biologist* (1985), is, in some ways, the pathway out of this epistemological dilemma. Yet it would be fair to say that all historical actors probably had something of a common understanding. From Darwin's "tangled bank" to Levins and Lewontin's "dialectical biology," all have been in agreement that complexity in one form or another, is a characteristic of ecological systems.

To "understand": nine reasons for generalization

In this book we seek to present what we think are generalizations. It is somewhat unfortunate that much of what passes for "ecological knowledge" comes from casual sources – videos about nature, backpacking expeditions to observe "nature," and so forth. While the observation of a lion kill in an African savannah or the mass flowering of trees in the dry forests of Central America is certainly awe-inspiring, neither actually represents, by itself, any sort of ecological understanding. The word "understanding" is a very difficult word philosophically, despite the fact that it is used quite frequently in everyday conversation. In a scientific sense the word is best viewed as derived from, or at least related to, the word "generalization," something true of ecology as much as of chemistry or physics. We can observe that weeds reduce crop yields and that wild buffalo reduce forage for pastoralists, and similar recurrent events and come to predict that weeds will always reduce crop yields and buffalo will always reduce opportunities for pastoralists, and perhaps that is all that is necessary. Our repeated experiences allow us to make sense of the world. However, once we generalize that both observations are examples of the same general thing, ecological competition, we have come to increase our "understanding" of nature, just a little. It is the generalization that creates, at least the feeling of, understanding.

Different sciences have different cultures of generalizing and understanding, indeed, in any "way of knowing," whether religious, scientific or otherwise, the nature of generalizing and understanding are variable and change over the course of time. In the particular case of ecology, we propose that there are nine precepts that provide us with insights into what ecological generalization and concomitant understanding, is about, or should be about, in the first place.

1   To repeat, the *key reason for generalization* in ecology, or, more simply put, ecological theory, *is that it allows for understanding*. Repeated observation may lead to future expectations, but only theory can lead to understanding. I may observe, year after year, that when my crops are planted in the shade of a tree they grow less rapidly than when planted in the open sun. And, because of this experience alone, I may expect that plants in general will grow more slowly when in the shade than when in the sun. Yet once I learn of the "theory" of photosynthesis, I "understand" why I have made that observation all my life.

2   Predictability is not the ultimate measure of understanding and *it is understanding, not simple predictability, that concerns us*. Repeated observation of a correlation between two phenomena yields the expectation (i.e., the prediction) that those phenomena are likely to be correlated in the future also, and indeed those predictions may be very accurate. But both

non-Western and Western intellectual traditions at their core, seek understanding, not predictability. Understanding is the goal of science; predictability is the goal of engineering.

3   Popper distinguishes between a bucket into which observations of nature are placed versus a searchlight that both illuminates observations and possibly finds previously obscured observations that then become illuminated.[21] The bucket metaphor of knowledge suggests that observations are repeatedly stuffed into a bucket and a measure of understanding is the number of observations contained in the bucket. The searchlight metaphor suggests that an underlying framework provides us with an understanding of observations made; expectations derive from a set of rules that produce them, not from the simple observations that they are in the bucket. In this context, theory is more like a searchlight. Certainly *filling our buckets with observations is an essential process* in science. It makes possible inevitable mental constructs without which many parts of the world would be so unpredictable so as to make even elementary movement impossible. *But our searchlights are what provide us with understanding.*

4   The relationship between theory and materialism is central to understanding both modern science and traditional knowledge. A theory based on idealistic conceptualizations is not scientific – the thing we see is real and material, not a manifestation of some ideal version of that thing. In this context, as noted by Levins and Lewontin in *The Dialectical Biologist*, it is important to distinguish between the abstract and the ideal. *Abstraction is necessary to formulate a platform of understanding* (a theory), *but idealism,* presuming that there is an ideal form for every observable, *leads to reductionism.* Platonic *idealism has no place in modern science* and, we argue, when allowed in traditional knowledge, leaves that knowledge deprived of deep understanding.

5   The formalities of induction versus deduction in developing theory are key elements that dialectically unite traditional knowledge, most often derived from a long history of induction, with Western scientific knowledge, most often derived from deductive reasoning. Going back to the example of plants growing in the shade, through experience, a farmer may conclude through induction that crops don't grow well in the shade, which is certainly sufficient for practice. The scientific approach is to seek a generalization and deduce particular circumstances from it. Farmer and scientist alike may arrive at the same conclusion but through different processes.

6   *All theory is ultimately metaphor or simile.* If the phrase "my love for you is like a ship lost on a stormy sea" provides you with some understanding of my feelings for you and how those feelings affect me, it is because you can visualize that lost ship and that stormy sea, tie that visualization to emotions, and thus gain understanding of how I feel. Scientific theory is identical. For example, an ecological community can be likened to a spider web with many connections that give it strength and stability or to a

house of cards balancing precariously on a few cards (e.g., keystone species). Each of those metaphors suggests an understanding of what provides structure to that community (of course any particular "understanding" could ultimately turn out to be misleading or wrong and in this sense, any understanding is provisional).

7   *A mathematical representation of theory is just a special kind of metaphor.* We effectively say "my love for you is like this equation." The equation is the metaphor that provides the understanding. One might argue that mathematical metaphors are in some way more important than others, simply because there is no question that we all understand them in the same way. The ship on the stormy ocean may unleash different emotions in a fisherman familiar with shipwrecks than in a shepherd who calmly watches the ships from the shore, but the stable equilibrium point of a system of equations is the same for all. Thus, although theory is not necessarily mathematical, formulating theory in a mathematical way, when possible, is far more convenient than other sorts of metaphors.

8   Concerning theory and prediction, the purpose of theory, mathematical or otherwise, is to provide a searchlight that illuminates our understanding. *Predicting a phenomenon from that theory is not the purpose of the theory; its purpose is understanding.* To be sure, prediction is part of the scientific process, but only as a way to measure the validity of the theory. Prediction in and of itself is not part of science. It is only part of the evaluative process of deciding whether a theory is useful or not.

9   *Our intuition about nature is ever-evolving.* Every day more experiences feed our buckets of observations, and our intuitions about how the world works are ever changing in the face of this data. However, theories are, in the end, another type of experience, they are searchlights that make every observation take on meaning that it would not have had without the theory. The fact that I can predict that the crops will grow more slowly under the dense shade of a tree is not an important part of the theory of photosynthesis. But the theory of photosynthesis gives me understanding of why the tree's shade reduces the growth of the crops.

A note on theory and intuition

Theory, especially its mathematical form, is regarded as a foundation stone for many sciences. It is at least a tool of perhaps all sciences, even if not foundational. Its utility is hardly questionable for the engineering sciences while its worship and perhaps overextension in economics and ecology has been amply criticized. Relying on complex models to understand complex phenomena can be problematic to be sure. When a model becomes synecdoche for reality, and reification creates seemingly real categories that fade rapidly upon close analysis, anything approaching "understanding" is obscured. As a colleague once noted, you can spend years developing a mathematical model of an ecological process that you do not understand, only to wind up with a mathematical model that you do not understand.[22]

When both the variables and parameters of a model or theory are well-defined, such that there is no question about either the meaning of the variables or the way they interact, a theory is a powerful thing that may very well allow for the prediction of important real things. Yet concepts like marginal utility or population density are simply not the same as force or atomic weight. They are convenient abstractions of things we see, or think we see, in nature. However, likening them to phenomena like force and atomic weight is reification – convincing oneself that they are more real than they actually are.

Despite the unfortunate excessiveness of the postmodern movement, the notion that we socially construct things was an important contribution to our understanding of how science proceeds. While there can be no doubt that theories, including those formulated mathematically, are social constructions, it is frequently the case that the reality they seek to describe is itself socially constructed. There is consequently a legitimate criticism of theory more generally. Yet we are not in a position to suggest that knowledge can be gained completely independently from theory – whether we like it or not, whether we find it difficult or easy, theory is the foundation of science both natural and social, and trying to understand the world in its absence is trying to fight with one hand tied behind our back.

While our presentation of ecological complexity in this book is an attempt at didactic simplicity, sometimes oversimplifying to what will probably be the disdain of more purist theoreticians, everything we talk about has a strong theoretical background supporting it. Furthermore, it must be understood that theory in ecology has become based, to a large degree, on mathematics. Nevertheless, it is our contention that the formal theory, mainly mathematical in origin, can be finessed since its significance can be explained with the simple tools of graphs and examples. That is, the subject matter presented in this book, even though it originates from theoretical and mainly mathematical thought, can be explained in a broad, qualitative way. This is not really all that unusual. Theory, especially when taking the form of mathematical models, has a utility that, even though historically common, is rarely acknowledged for what it is. Richard Levins referred to this form of utility as "educating the intuition with mathematics," a position that informs this book at its foundation.

Ecological complexity and the organization of this book

In the modern world, the science of complexity, anticipated by Rachel Carson and long advocated as the proper foundation for ecology by Levins, can form the basis of an ecological theory that eschews idealistic notions yet acknowledges the truth of the whole. As we noted earlier, complexity in its modern scientific sense, either the dialectical materialism of Levins and Lewontin or the essentially equivalent "complex systems" of Miller and Page and other more recent complexity scientists, should not be confused with "complicated." It is an easy trope to note that ecological systems are complex because they include many interconnected components. But complexity in its modern sense is something different. As we noted earlier, in complex systems we have two fundamental examples that frame the subject: first, that simple things can generate complicated patterns (Figure 1.2a), and second, that complicated things can generate simple patterns (Figure 1.2b). It is in the spirit of the wonders of science that we note the first one. For example, three planets, obeying the simple rules of Newtonian mechanics, produce patterns that are 100% non-predictable, as was anticipated in the nineteenth century by the French mathematician Poincaré. It is in the spirit of the deep insights that can be provided by science that we note the second one. For example, virtually unlimited numbers of species, interacting in unknown ways, produce a predictable species-to-area relationship. Complex systems are that way. And ecological systems are examples of complex systems.

As we apply this framework, we need to remain cognizant of the fact that the science of complexity is evolving, as all good science should be.[23] Thus, what we present in this book are the components of complexity science, as we know them today, that we think are relevant to agroecology. Some readers may notice that this approach rejects previous popular notions about nature, such as the idea that biodiversity leads to stability, or that there is a "balance of nature," or that most natural ecological communities are at equilibrium. Most modern ecologists agree that such ideas about nature are anachronistic. Yet some of them took strong hold on the popular culture. Likewise, some of the concepts related to ecological complexity may prove to be wrong with time.

For this book, we have identified seven topics that together, or the intersection of which, we feel encapsulate the current concept of ecological complexity, namely, (1) Turing processes and spatial structure, (2) chaotic dynamics, (3) stochastic processes, (4) coupled oscillators, (5) multidimensionality, (6) trait-mediated indirect interactions, and (7) critical transitions. Few contemporary ecologists would deny the importance of any of these, yet may not go as far as we wish to go in suggesting that they all combine into one qualitative whole. Others will note that we fail to include other important topics. Both may be correct, but this is our vision, based on our experience working in agroecosystems, of how concepts of ecological complexity can enlighten the science and practice of agroecology. We suggest that it is the interpenetration of these subjects together that constitutes what should be called ecological complexity. A pictorial representation of this idea is presented in Figure 1.3.

This book is organized according to the conceptualization in Figure 1.3, with seven distinct chapters, preceded by this introductory chapter and Chapter 2, which introduces some of the very basic ideas upon which the rest of the chapters rely. Readers familiar with the basic concepts of terrestrial ecology can skip Chapter 2. Finally, in Chapter 10, we summarize it all with a vision of how further development of the science of ecology may enhance, and be enhanced by, the growing scientific/political movement called agroecology.

Figure 1.3 Conceptualizing ecological complexity as the intersection of the seven overlapping topics thought to be important in the field of agroecology (image provided by Theresa Ong)

Notes

1 An enormous literature now exists making this point in one way or another. Exemplary book length examples that have proven useful to us are Oelhaf, 1978; Jackson, 1980; Altieri, 1987; Carroll et al., 1990; Pretty, 1995; Conway and Barbier, 1990; Jackson, 2002; Rickerl and Francis, 2004a; Gliessman, 2006; Bohlen and House, 2009; Vandermeer, 2011; Goodman et al., 2012.
2 Vasey, 2002; Heckman, 2006; Harwood, 1990.
3 Rosset and Altieri, 1997; Rosset, 2013; Francis et al., 2011; Pretty and Buck, 2002; Piper, 1999.
4 Egerton, 2011.
5 Mitchell, 2009; Holland, 2012.
6 Complex systems is emerging as a legitimate scientific discipline in and of itself. Recent texts include Johnson, 2009; Mitchell, 2009; Miller and Page, 2007; Holland, 2012, 2014; Byrne and Callaghan, 2013.
7 Graph is from figure 2 on page 8 of "The theory of island biogeography" by MacArthur and Wilson, based on research by Darlington circa 1957.
8 Levins and Lewontin, 1985, p. 3.

9   Miller and Page, 2007.
10 While it is difficult to come up with a truly accurate number, what is clear is that a vast proportion of the land devoted to industrial-style agriculture occupies a vast majority of land currently categorized as agricultural land, while at the same time the vast majority of farmers are small scale. Relevant data are presented in the International Assessment of Agricultural Science and Technology for Development (IAAST) report, and elsewhere.
11 McLoone, 1998.
12 Howard, 1953; Travers and Milgram, 1967.
13 Stanhill, 1990; Balfour, 1982.
14 Campbell, 2009; Loewenstein, 2001.
15 Cafaro, 2013; Lear, 1998.
16 Hecht, 2010; Van der Ploeg, 2009.
17 Desmarais, 2012; Rosset, 2013.
18 The quote is from the introduction to the 1943 edition of "An Agricultural Testament", available at http://gutenberg.net.au/ebooks02/0200301.txt. Italics in brackets are our extensions of his original statement.
19 Simberloff, 1980.
20 Bellamy Foster et al., 2010.
21 Ter Hark, 2009.
22 As Levins once quipped, "The most precise model of a beaver is another beaver – preferably the same one."
23 Although our approach is influenced by many authors, Dick Levins obviously stands out as our main influence. Others would undoubtedly have cited other main influences since it is arguably the case that many other authors have, almost independently, emphasized the complex nature of the natural world, if not specifically agriculture. Other popular analysts include the frequently elegant work of Belgian physical chemist Ilya Prigogine, especially the work coauthored with Isabelle Stengers (1997). Far more complicated is a deep mathematical definition of complexity by Robert Rosen (recently dubbed Rosenennian complexity), which we recently commented on (2017).

2 Basic ecological concepts

Before developing the core ideas of ecological complexity, we introduce what we regard as the key basic ecological concepts relevant to agroecosystems. This chapter is meant for those who do not have a background in ecology or who wish to review some basic ecological principles before proceeding with the rest of the book. Note that our coverage is highly biased, incomplete and aimed at only those subjects that we think are relevant for the subject matter covered in the rest of this book – many other important subjects in ecology are normally covered in both ecology texts and elementary courses.[1] Furthermore, the science of ecology today (writing in 2017) seems to be divided into two distinct categories: ecosystem ecology and population/community ecology. Such a division has come about to some extent because of the intellectual questions posed, but also because of the methodologies involved. For purposes of this text we feel that the distinction is unimportant. Complexity, as conceived here, applies to both wings of the ecological community. Thus, in this introductory chapter we simply avoid the use of the terms, although it will be obvious to most that some of our subjects are more commonly thought of as ecosystem ecology, while others are clearly in the realm of population/community ecology. The nine basic subjects are:

1 How plants get energy
2 How plants get nutrients from soil
3 Transformation of nutrients in the soil
4 Water–soil–plant interactions
5 The ecological niche: a historical backbone
6 Population dynamics
7 Equilibrium, resilience, persistence
8 Basic trophic dynamics
9 Feedback in dynamic systems.

How plants get energy

It is general knowledge that plants are extremely different than animals in that their energy-gathering organs are distinct from their material-gathering organs. We humans, as all other animals, gain energy from the food we eat, pure and simple. We digest food and from that digestion we manage to break energy-rich chemical bonds, releasing the energy they contain so as to do useful (sometimes) work with it. At the same time we use the materials that are bonded together by those chemical bonds to make ourselves – our proteins, our carbohydrates, our fats. Some of our parts we store and use later (or store permanently until we die – another story). There is much to be said generally about what and how we eat, but the point here is that the two things we need to live, energy and material, are obtained by the same system of organs, our digestive system.

Plants, on the other hand, have one system (leaves) for gaining energy (through the process of photosynthesis) and a different system (roots) for gaining materials (through nutrient absorption from the soil). For us, the old adage "you are what you eat" makes a lot of sense. For plants it would have to be "you are what you, on the one hand, photosynthesize and, on the other hand, what you absorb from the soil" – not terribly catchy.

It is worth reflecting for a moment on what this basic difference implies. We, animals in general, are certainly not necessary for life on earth. At least theoretically one could imagine a life system in which all organisms are autotrophs (make their own energy) – actually the earth was probably this way a few billion years ago. We heterotrophs are energy parasites since we cannot, independently of other life forms, gain our needed energy on our own. Yet, at our present level of evolutionary development, from an ecological point of view it would be clearly impossible for the autotrophs to take over. Current terrestrial ecosystems have a general structure in which plants suck energy out of the sun, herbivores gain their energy from ingesting plant material, and carnivores gain their energy from ingesting animal material. There would be no problem at all imagining a terrestrial ecosystem with no herbivores or carnivores (indeed the plants might be quite happy about that). But plants also have that other fundamental life system, the absorption of the materials necessary for constructing the parts that permit them to suck out energy – the processes that go on under the ground. And these processes are not autotrophic in the same sense as is photosynthesis. There are animals (and bacteria, fungi and archaea[2]) that are intimately involved with this process, and, at least in their current form, the world's main autotrophs in the terrestrial world would not survive without the bacteria, fungi and archaea that live in the soil and produce the conditions whereby the materials needed by the plant are made available to it. So the truly autotrophic process, the process that gains energy from the sun, is separate from the other function, obtaining construction materials, which is, at least in formal definition, heterotrophic. We follow tradition and speak of them separately, first photosynthesis and second nutrient capture, but insist that the reader note their essential interdependence.

The basic ecological processes involved in photosynthesis are well-understood. Solar radiation reaches the leaf in the form of photons, which are effectively massless packets of energy. They are what makes the sunlight feel warm as it strikes your skin. A single proton strikes a molecule of chlorophyll, effectively knocking one of the electrons in part of that molecule into an "excited" state. That excited state means that the electron, for a very short period of time, contains an extra amount of energy (that it got from the photon). If nothing else happens, that energy will dissipate into the environment as heat, the same heat that the photon originally carried. But something does happen. There is an elaborate process contained within the photosynthetic chemical apparatus that captures that energy from the excited form to a form that can be stored and used later. The chemical details of that process are well-known, but not especially relevant for understanding photosynthesis from the point of view of agroecosystems. There are, however, two relevant issues: (1) the nonlinear form of energy capture with respect to sunlight and (2) the dual form of energy capture across the plant kingdom (at least that part of the plant kingdom that includes most of the world's crops).

A graph of energy gained as a function of the amount of incident light generally takes the form of an increasing curve with diminishing returns, as pictured in Figure 2.1. Naturally, to capture the energy it is necessary to expend some energy, so, as in most of life's situations, we contrast gross revenues (amount of energy absorbed) to net revenues (amount of energy absorbed minus the amount of energy expended to store it). The amount of energy expended includes a substantial amount that needs to be invested in maintaining the plant material already there, so it makes sense to think in terms of "negative" energy, that is, the energy needed to maintain the plant even in the absence of any light whatsoever. These ideas are also presented in Figure 2.1.

In viewing Figure 2.1 it makes sense to use a bit of shorthand and say that light is a "resource" that a plant needs and that the plant "consumes" the light. As we examine the physical/chemical nature of photosynthesis the words consumer and resource may not be the best descriptors of what happens, but viewing the process this way brings us to one of the most important ideas in ecology, the consumption of resources. It emerges again and again in many contexts, so we employ it here to describe the way plants respond to light.

The process of photosynthesis is actually quite amazing. Imagine trying to describe it from simple day-to-day observations, which is to say, could you understand it from the process of pure induction? Yes, farmers the world over generally understand that crops need sunlight to grow. But how would you infer from that knowledge that it was carbon dioxide (the stuff that animals breathe

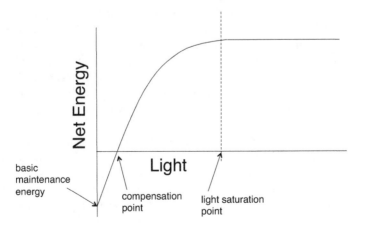

Figure 2.1 Light response curve in photosynthesis

out) in the air that the plants use to make carbon structures that eventually turn into their bodies? Indeed one of the first truly groundbreaking experiments actually concluded the wrong thing. Justus von Humboldt, in 1648, planted a 5-pound willow tree in a pot containing 200 pounds of soil. He did nothing else except add water to the plant regularly. When he harvested the tree 5 years later the tree weighed 169 pounds but the soil lost only 2 ounces during that 7 years. His conclusion, as in all intellectual activity, was limited by his background assumptions about the world which, in this case, was that the only possible sources of the physical reality of the plant was either the soil or the water. Since the soil weighed virtually the same before and after the experiment, it must have been, he reasoned, the water that provided the material that the plant used for growth. Our understanding that it was the carbon dioxide in the air would have been difficult for him to grasp, given the general state of knowledge in 1648.

Neither this historical note nor the chemical details of photosynthesis are all that important for ecological applications, but the fact that two major forms of photosynthesis have evolved is of some importance to agriculture in general, and certainly to many aspects of the ecology of agroecosystems. After the capture of energy by the chlorophyll molecule, the excited (energized) electron loses its additional energy quite rapidly, so for that energy to eventually be used by the plant it needs to be captured by some other processes. Although the details of this capture are fascinating in and of themselves, they are not crucial for understanding the rest of the process. Just know that there is a chemical apparatus that catches most of that energy before it escapes as heat into the environment. The chemical that first incorporates that energy (eventually giving it up to other chemical processes) has a carbon backbone, and curiously there are two alternatives for the initial energy capture. The first uses a backbone composed of a three-carbon molecule and the second uses a backbone composed of a four-carbon molecule. Thus we refer to the two types as C3 and C4 photosynthesis.

In Figure 2.2 we illustrate the difference between a C3 and a C4 plant. The most significant ecological issue of these two modes is the pattern of energy accumulation as a function of light. C4 plants have a lower compensation point than C3 plants, and usually a higher saturation point, enabling them to take advantage of higher sunlight intensity to produce more. Many C4 plants are tropical grasses, and some of those tropical grasses have been domesticated (e.g., maize, sugarcane, millet, sorghum), but the majority of crop species are C3.

An additional pattern of light capture in photosynthesis, recognized for centuries, has to do with the response of individual plants to different levels of light. Frequently an approximate classification is made between those plants that do well in full sunlight (called "sun" plants, or "shade-intolerant" plants) versus those that are capable of photosynthesis at relatively low light intensities (called "shade" plants, or "shade-tolerant" plants). These categories are frequently important elements in traditional agroecosystem planning. For example, agroforestry systems around the globe tend to combine sun plants (e.g., an overstory of fruit trees) with shade plants (e.g., shade-tolerant root crops), in a system that, for fairly obvious reasons, more fully exploits the available light.

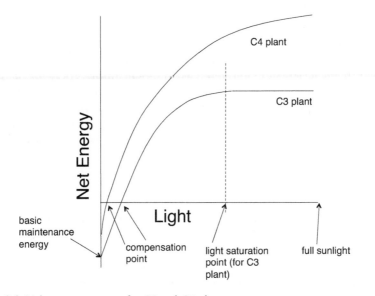

Figure 2.2 Light response curves for C3 and C4 plants

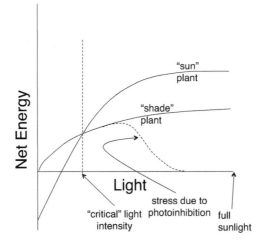

Figure 2.3 The basic difference in light response curves between sun plants and shade plants, and the the effect of photoinhibition

Although beyond the intended scope of this chapter, there are certain bio-chemical interactions that sometimes imply a relationship between shade toler-ance and photoinhibition – the reduction in potential photosynthesis at high levels of light availability, as illustrated in Figure 2.3. Again, the implications of this process for agroecosystem planning are obvious.

How plants get nutrients from soil

The other part of the plant's life apparatus is the root system, whereby the essential materials for life are obtained. As we noted earlier, the root system is actually part of a complex community of organisms that live in the soil, process the material in the soil, and make the elements necessary for the plant available. In this simplified account we note that there are effectively three functional groups of microorganisms involved: (1) nitrogen-fixing microorganisms, (2) decomposers, and (3) chemical catalyzing microorganisms. The third category contains several important subgroups, which we will discuss more completely.

Nitrogen-fixing microorganism

Nitrogen-fixing microorganisms are central to the agroecosystem. Much of the structure of life is made of proteins and much of the coordination of life is orchestrated through enzymes, which are also proteins. One of the major constituents of protein is nitrogen (N), giving that particular element a key role in life's processes. However, especially in the agroecosystem, nitrogen is continually removed from the system through harvest. Indeed, Marx and Engels talked about a "metabolic rift"[3] in the sense that nutrients are pulled out of the soil in agriculture, harvested in the form of food, transported to the cities, reprocessed (eaten), and dumped into rivers. The natural ecological metabolism is thus dramatically altered in the agroecosystem because of this division between town and countryside. The nitrogen in the fields is used to make the rice that is transported to the city, but the wastes (human and otherwise) from that transported rice, heavy with nitrogen, are discarded into the rivers and eventually into the ocean. Effectively, the materials in the soil are being mined and transferred to the ocean. This metabolic rift is actually not much of a problem when considering nitrogen (it is, however, a big problem for phosphorous and other elements) because we have an almost limitless supply of nitrogen in the air around us. However, that limitless supply of nitrogen is made available through the activity of microorganisms, that take the nitrogen molecule in the form of gas (N_2) and embed it in amino acids and proteins in their bodies, a process called nitrogen fixation.

From an agricultural perspective, most of these nitrogen-fixing microorganisms are bacteria that live within special organs called nodules in specific plants, most usually legumes. These nodules are specialized structures that plants make specifically to house the nitrogen-fixing bacteria, providing the latter with housing and reaping the benefits from the proteins the bacteria are able to make from the nitrogen in the air. From an ecological perspective, nitrogen is effectively "pumped" into the ecosystem through the action of nitrogen-fixing bacteria. In ecology these mutually beneficial interactions are called mutualisms. This particular kind of mutualism, where one of the partners lives within the other, is called symbiosis.

Decomposers

Non-leguminous plants generally do not have the help of nitrogen-fixing bacteria and must absorb nitrogen in the form of nitrate or ammonium (recall that the first is an anion, or negatively charged ion, and the second a cation, or positively charged ion). How those two elements are made available to the plant is a complicated story that involves several basic forms of microorganisms. First among them is a group of organisms that specialize in eating dead (organic) material – the decomposers. This functional group includes a wide variety of organisms, from earthworms to bacteria, from millipedes to fungi. The decomposition process is inevitably complicated and highly variable from place to place. But from an ecological point of view we see highly structured organic material gradually decompose into ever smaller units, first by the action of large organisms like earthworms and millipedes that provide shredded material in their feces upon which fungi and bacteria feast, using the chemical bonds within the organic material for their own energy supply, eventually resulting in molecules small enough to be absorbed by plant roots.

Since amino acids form the backbone of proteins and since amino acids have ammonium as a basic structural component, it is not surprising that the first ionic product resulting from the decomposition process is ammonium. If not absorbed directly, the ammonium, which is a cation, is converted to nitrate, an anion. This conversion is facilitated by a specialized group of bacteria called nitrifying bacteria (one of the categories we discuss under "Chemical Catalizing Microorganisms"). Plants generally are capable of absorbing both ammonium and nitrate, but the former has to be stored in special vacuoles and must be used rather quickly, since it has detrimental effects on cells if its concentration gets too high. However, from a fundamental energy argument, since ammonium is a building block for amino acids which are the building blocks for proteins, plants that absorb ammonium from the soil save on the energy needed to construct ammonium from nitrogen (as is necessary when nitrate is the absorbed ion). Furthermore, being a cation, ammonium is held in storage in most soils while nitrate is an anion and consequently "leaks" out of most soils quite rapidly. In a sense ammonium represents a slow-release nitrogen source, while nitrate must be taken up rapidly or it will leach out of the system.

The underlying cause of leakage or storage is mainly determined by the colloidal structure of the soil (colloids are particles so small they will not settle out of solution even when the solution is perfectly still). There are actually many different forms of colloids but the main feature that concerns us here is their electrical structure. They contain on their surface an abundance of negative electrical charges (they also contain positive charges, but the vast majority are negative). From basic physical principles, the cations and anions in the solution surrounding a colloidal particle will be representatively attached to the colloidal surface in proportion to their relative concentrations in the soil solution surrounding that particle. So, for example, if the soil solution contains 80% ammonium, 10% phosphate and 10% potassium, the cations that attach to the colloidal surface will be in precisely those same proportions (80% ammonium, 10% phosphate and 10% potassium).

Near to the surface of a root hair (recall that a root hair is an evagination of a single cell in the cortex of the root), the plant pumps carbonic acid and hydrogen ions into the soil solution. Both are cations. This means that the relative concentrations of various cations in the soil solution will tend to be quite different in the vicinity of a root hair or the rhizosphere. There will be extra positive ions in the form of hydrogen (and carbonic acid, among other things that are pumped out of the root hair) in the solution in that vicinity. So as to balance the relative concentrations on the surface of the colloidal particles with the surrounding soil solution, other cations (like ammonium, calcium and potassium) have to be released from those colloidal surfaces. As these other cations are liberated from the colloidal surfaces they are made available to be absorbed by the plant through the root hair. This "pumping" action at the surface of the root hair creates a sort of overall flow of cations toward the surface of the root hair.

Note the way the relative concentrations of potassium and hydrogen are pictured in Figure 2.4. Far away from the root surface the concentration in both the soil solution and on the surface of the colloids is about 10 potassiums to 3 hydrogens. Near the surface of the root, the concentrations are about 7 potassiums to 8 hydrogens – the surface potassiums in the soil solution far from the

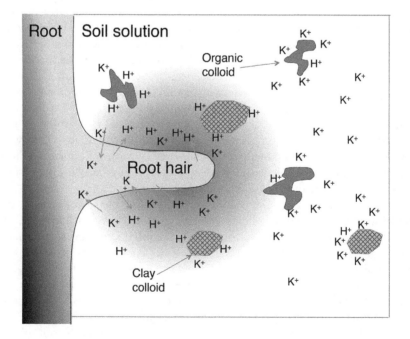

Figure 2.4 Basic process of cation exchange in plant nutrition, illustrated with potassium dynamics; color gradient bubble surrounding the root hair represents the concentration gradient of potassium

root have been exchanged with surface hydrogens when the colloid comes near to the root surface. Thus, many of the cations that were on the colloidal surface in the general soil solution have to be released into the solution to make room for the hydrogen and carbonic acid ions that have been released from the root hair. Cations like ammonium, calcium and potassium are released from their attachment to the surface of the colloids so as to restore the balance between the soil solution and the colloidal surface. The final consequence is that these cations, having been exchanged on the surface of the colloid, are now in solution and can be directly absorbed by the plant through the root hair. The general process is illustrated in Figure 2.4.

Chemical catalyzing microorganisms

A variety of other microorganisms that live in soils are also important for nutrient acquisition by plants. As mentioned previously, a group of bacteria called nitrifying bacteria are responsible for transforming the ammonium cation into a nitrate anion. This process can be important in agriculture because nitrate, as an anion, is more easily leached out of soils than ammonium. Although there is an equivalent process of anion exchange, it is trivial in most agricultural applications since the net charge on most soil colloids is overwhelmingly negative. Lacking the absorptive attraction of the colloids (because of the negative charges), there is nothing "holding" the nitrate in the soil, meaning that it leaches out very easily.

Another group of microorganisms reverse the process of both the nitrogen fixation and nitrification by turning nitrate into gaseous nitrogen or nitrous oxides, a highly potent greenhouse gas. These microorganisms are called denitrifying bacteria and can be very important for agriculture because they can lead to major nitrogen losses from agricultural soils, especially if excessive fertilizer is applied.

A third group of microorganisms important for nutrient acquisition in plants are the so-called phosphorus solubilizing microorganisms: bacteria and, especially, fungi. These microorganisms are capable of solubilizing inorganic phosphorus from insoluble compounds. It is thought that the way they do this is through the release of organic acids that chelate the cations bound to phosphate, converting it into a soluble form that can be absorbed by plants. Phosphorus is an important macronutrient for plants since it plays a role in photosynthesis, respiration and energy storage among other essential processes for the life of plants. Many types of soils have high concentrations of phosphorous but they are usually chemically bound either in recalcitrant molecules or in the bedrock and not accessible to plants. Phosphorus solubilizing bacteria release this important element making it available for plant absorption.

Transformation of nutrients in the soil

Pictured in Figure 2.4 are two general types of colloids, organic colloids and clay colloids. Both function in the same way with respect to the exchange of cations in the rhizosphere. But this brings up an extremely important issue for the ecology of soils. The organic colloids come from the decomposition process. Humus, commonly thought of as providing the black coloration to well-prepared compost, is an organic soil colloid. It too is brought under continual biological activity by decomposing microorganisms, but its chemical structure makes it difficult for those microorganisms to break it apart and thus it remains as a soil colloid for many years – it is what soil scientists refer to as recalcitrant. It is best to think of organic colloids as constantly changing, but at a very slow rate. The pool of organic colloids is continually *replenished* by the decomposition process acting "upstream" but continually *depleted* by the decomposition process acting "downstream." Yet, as mentioned earlier, the decomposition process itself is responsible for releasing the small molecules that the plant is then able to absorb through its roots.

The best way of viewing the whole process of decomposition is as a dual process. When a piece of organic material enters the soil system (say, a leaf dies and begins decomposing), there are two distinct functional pathways that the process takes, depending on the micro parts of the leaf and the local surrounding environment. With the part of the leaf composed of molecules that are relatively recalcitrant, the decomposition process is very slow, operating over tens or even hundreds of years. The rest of the leaf is composed of material that decomposes rapidly, releasing, usually in about a year, the small molecules that the living plants can absorb. Thus there are two decomposition cycles: the fast cycle, which releases essential nutrients to the soil solution; and the slow cycle, which creates the organic colloids necessary for cation exchange. This means that the fundamental ecological process of decomposition involves, first, feeding the plants by making small molecules available for the plant to take up, and second, making the soil healthy for storing and providing some of those molecules by continually feeding organic colloids into the soil solution.

In the past it has been customary to view the decomposition rates of different organic compounds as mainly a property of the chemical structure of the compounds themselves, as we did in the previous two paragraphs. Recently this view has been challenged. In a recent article in the journal *Nature*, for example, a team of soil scientists noted:

> Rather than describing organic matter by decay rate, pool, stability of level of "recalcitrance" – as if these were properties of the compounds themselves – organic matter should be described by quantifiable environmental characteristics governing stabilization, such as solubility, molecular size and functionalization.[4]

In other words, environmental factors interact with the organic material to produce breakdown products, some of which last for long periods of time while others decompose rapidly. Although this way of looking at the world is probably more precise, it remains the case that some materials wind up persisting in the soil for many years and effectively contribute to soil physical structure, while others decompose very rapidly and provide the nutrients that are needed for plant growth and the carbon needed for microorganism growth.

Although the plant absorbs a variety of different micro molecules from the soil, one of those molecules is frequently the most critical of all the critical elements — nitrogen (the elements required in large numbers are, more or less in this order, nitrogen, phosphorous, potassium, and calcium). From the point of view of the growing plant (or the farmer), the nitrogen cycle is a very dynamic process, and making sure that the nitrogen supply is always adequate for plant growth is challenging for the farmer and, metaphorically, for the plant too. Nitrogen is fixed biologically at a rate partially dependent on the biomass of the plant involved in the fixation process. When the plant dies, its tissues are converted to organic matter by the process of decomposition, producing, in part, ammonium, which is taken up by the plant roots.

One of the most important things to realize about this process is its fundamentally non-linear nature. In Figure 2.5 we illustrate the most basic elements of that process. In Figure 2.5a, we see the death of a living plant (e.g., a crop) giving rise to dead plant material which is converted, by the action of a host of microorganisms, into ammonium and then nitrate, which is absorbed by the still-living plants in the system. The most important general feature to note is the dual role of microorganisms, first as a user of nitrate and second as a promoter of more nitrate being produced. We have made the diagram so as to emphasize the key dynamic feature of the microorganisms as an indirect mediator of the rate of transformation of dead plant material to nitrate, by the arrow from the microorganisms to the arrow connecting dead plant material to nitrate. This way of presenting the information emphasizes the role of an essential non-linear process in the system (the microorganisms directly affect not the dead plant material itself, but the rate at which that material is transformed into nitrate). As we elaborate a bit more fully in the following paragraphs, the overall system is far more complicated than what we present in Figure 2.5a, but our intent here is to note that at its most simplified level, there is a key, and unavoidable, nonlinearity in the system. Once one begins adding nonlinear terms to a dynamic process, all sorts of complicated things happen, and those beautiful idealized notions of equilibrium and balance begin to fade into their proper place as anachronistic desiderata.

In Figure 2.5b we present this same story in a slightly more mathematical form and highlight how the four fundamental dynamic processes are connected to one another with four pseudo equations. Most important is the fact that there is a nonlinear term in the first and second equation, "rate of input from living plants" is *multiplied* by "metabolic rate of microorganisms." It is important to appreciate that these sorts of nonlinear terms are incredibly important in ecology, indeed sometimes, if not usually, dominate the dynamical structure of ecosystems.

As represented in Figure 2.5, the fate of nitrate is dual: it is absorbed either by the microorganisms (which, in soil science parlance, is frequently referred to as "immobilized") or by the plants. In reality there are four fates: two were already mentioned, but nitrate can also leach out of the system and it can be transformed into nitrous oxide, a gas which escapes to the atmosphere. When leached out

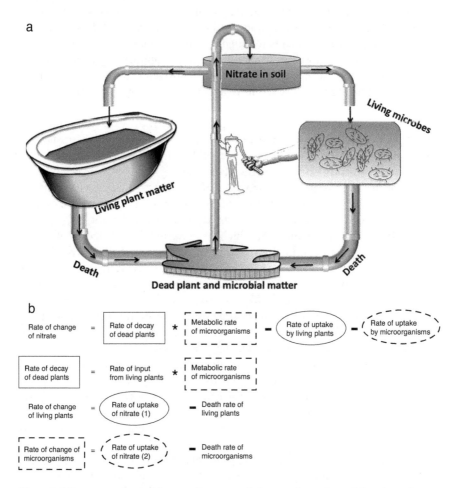

Figure 2.5 Representations of the overall process of nitrogen dynamics. a. Metaphor of water pumping in pipes. b. Mathematical relationships of the processes illustrated in (a), where the various rates are outlined in different styles to indicate where the same rate fits into each of the equations.

of the system, the nitrate frequently winds up in ground water or runs off into local streams and rivers, eventually reaching the ocean, where it is partially responsible for the generation of so-called dead zones.[5] When transformed into nitrous oxide, it fills the air with this very effective greenhouse gas. Eventually, most of the nitrous oxide diffuses to the stratosphere where it is converted to molecular nitrogen. At the other end of the process, the transformation of dead plant material to nitrate also involves some key steps, from (1) dead plant material to (2) organic matter to (3) the cation ammonium to (4) the anion nitrate, all of which are dependent on the catalizing action of the microorganisms. This chain of transformation is illustrated in Figure 2.6.

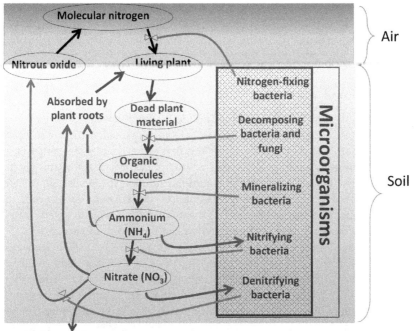

Figure 2.6 Simplified summary of the nitrogen cycle

The ammonium cations are adsorbed on colloidal surfaces in proportion to the other ions in the soil solution. At a rate dependent on the biomass of the appropriate microorganisms, the ammonium cations are transformed into nitrate anions and (1) absorbed by plants, (2) leached out of the system, or (3) denitrified to return to the atmosphere as nitrous oxide and eventually molecular nitrogen. In other words, it is sort of like filling a bucket having many leaks. The goal is to supply the nitrogen to the growing plant at exactly the correct rate, which implies controlling all the other leaks in the bucket, all the time, realizing that as the plant grows it changes its needs, and as the composition of the microbial community changes, the size and nature of the leaks in the bucket may also change. It is indeed a tremendously challenging problem for the plant and the farmer.

It is important to note here that all of these interactions are moderated to some extent by bacteria and fungi in the soil as explained in the previous section. Everything labeled "microorganisms" in Figure 2.6 is composed of carbon. Indeed, the chemicals making up the bodies of all these microorganisms represent a huge storage of carbon, required for making those chemicals. So, while the microorganisms respire and in so doing release CO_2 into the environment (sometimes methane and nitrous oxide too), the integrity of their bodies and the integrity of the bodies of their offspring rely on carbon that is integrated into body parts and functions.

Indeed, it is frequently noted that the limitation on growth of bacteria and fungi in the soil is carbon, not nitrogen, which is frequently the main limiting factor for the growth of plants. This is the reason it is frequently noted that soil is one of the biggest carbon sinks in the terrestrial world, and one of the reasons that agriculture is so important for the issue of global climate change.

Plants, soil and water

Note that in the previous section nothing much could have happened without water. The ions necessary for plant growth are all contained in the soil solution, which is water. And it is, we think, obvious that water is not a constant feature of the environment, coming and going with rainfall (or irrigation) and gravity in a complicated dance. Much of that complication has to do with supply and retention of water, which is strongly influenced by the pore structure of the soil – a complicated issue, but one that is generally dealt with in a very simple classification, under the formal names of soil texture and soil structure.

The measure of soil texture is, to some extent arbitrary, but is generally agreed upon by most soil scientists (and farmers and analysts). Considering all particles in the soil that are less than 2 mm in diameter, we can classify them in a tripartite classification: clay is 0–0.002 mm, silt is 0.002–0.05 mm, and sand is 0.05–2 mm. The relative proportion of each of these categories then is a measure of soil texture, and we typically categorize soils based on the relative proportions of these particles. Unfortunately the cutoff points for each of the particle sizes is different in different parts of the world, but the general principle is the same – if most of the particles are in the range of 0.05–2 mm, we say the soil is a sandy soil; if most are in the range of 0.002–0.05 mm, we say the soil is a silt soil; and if most are in the range of 0–0.002 mm, we say the soil is a clay soil. Soils that have a combination of all three in relatively similar proportions are called "loam" soils, frequently thought to be the "best" texture, especially with regard to water retention and percolation in agriculture.

All nutrient issues are associated with water, as we noted earlier, and water in soil is strongly affected by soil particle distribution, which is why soil texture is a fundamental property to be considered when concerned with water. Plants have a special role to play in the overall dynamics of an ecosystem, not just as primary producers, but as "end users" of water. Plants also face a contradiction dictated by physical conditions, as we illustrate with a simple metaphor in Figure 2.7. Plant roots suck the water up, and gravity pulls it down. The resistance is determined by the texture of the soil.

How rapidly the water changes in the soil is obviously affected by rainfall (or irrigation) and evaporation, but also by the two basic forces illustrated in Figure 2.7, the upward force from the pumping of the plant roots, and the downward force from gravity. Assessing the actual state of the soil at any point in time is generally important, but most especially when irrigation decisions must be made. Naturally in "swamp" production (e.g., paddy rice) or pure rain-fed agriculture, such a decision is not so critical, but in small-scale irrigation activities it is indeed an important feature. In Figure 2.8 we illustrate the basic categories used by some soil scientists to categorize the current state of a soil.

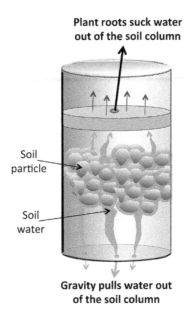

Plant roots suck water out of the soil column

Soil particle

Soil water

Gravity pulls water out of the soil column

Figure 2.7 Piston/cylinder metaphor of soil water dynamics. Soil is contained within the column (cylinder) and the plant sucks water out through the pumping action of the roots (represented by the arrow indicating that the piston is being pulled upward). Gravity represents a countervailing force in the system. The texture of the soil (relative size of the particles) is obviously a key element in resolving the contradiction between root pumping and gravity clearance.

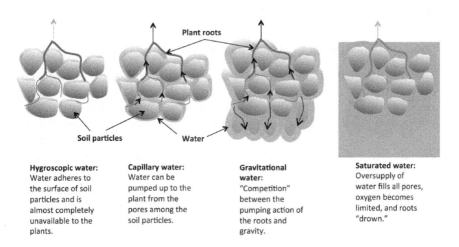

Plant roots

Soil particles   Water

Hygroscopic water: Water adheres to the surface of soil particles and is almost completely unavailable to the plants.

Capillary water: Water can be pumped up to the plant from the pores among the soil particles.

Gravitational water: "Competition" between the pumping action of the roots and gravity.

Saturated water: Oversupply of water fills all pores, oxygen becomes limited, and roots "drown."

Figure 2.8 Cartoon representation of the state of soil water

Especially with rain-fed agriculture there is a kind of competition between water availability and water storage. It is always the case that water tends to run out of the system due to gravity. And if a soil is composed of nothing more than sand, it is pretty clear that the very large pores in the soil will allow water to percolate very rapidly out of the soil, either gravitationally or through evaporation. Consequently, a sandy soil has very low water storage capacity. At the other extreme, a soil composed of nothing more than clay will have very narrow pore channels and thus neither gravity nor evaporation will remove water very rapidly (most will be held hygroscopically – see Figure 2.8). However, the plant roots will have to invest a great deal of energy in pumping the water out (imagine sucking on a very narrow straw). So a clay soil stores water efficiently, but the plants have a tough time getting it, while a sandy soil does not store water very well, but the plants can easily pump whatever they need while it is available. This is the basic contradiction that is partly resolved by some sort of loam soil, and is the reason that loam soils are generally regarded as "best" for agriculture. It is all about water.

Soil texture is the most easily conceptualized and measured quality of soil. However, it is not all that matters. Soil is far more than the distribution of particle sizes, and when trying to characterize soil, whether a traditional agriculturalist or a modern soil scientist, classification becomes a far more complicated subject. This more elaborate topic is usually referred to as soil "structure" and includes the final state of a soil, with all the organic matter that interacts with the mineral particles forming "aggregates," which affect water-holding capacity as well as availability to the plants. It is generally thought that having relatively large amounts of organic matter is beneficial to the plants. As we saw in the previous section, this benefit comes in part from the fact that organic matter acts as a cation-carrying colloid, but also because it can act as a particle that either holds water hygroscopically even when relatively large, or help generate larger pore spaces, resulting in more water loss through gravity and evaporation.

The ecological niche: a historical backbone

Perhaps no ecological concept has seen as much durability as the idea of an ecological niche. Despite a large amount of criticism of the concept,[6] mainly because it is difficult to define or measure precisely, both ecologists and farmers (as well as international political commentators) regularly use the concept, if not the actual term, niche. An organism somehow has a place in the world, and that place is somehow its ecological niche. Minimally as historical artifact, but probably more importantly as a main foundation for the science of ecology, the concept at least needs to be acknowledged. From the niche a plant occupies to photosynthesize, to the niche of bacteria promoting the transformation of ammonium to nitrate, to the niche of an herbivorous pest, to the niche of the predator of that herbivore, it is frequently useful to engage in "niche talk."

A plausible argument can be made that Darwin was the first to articulate the idea of the ecological niche, although he never used the term. On the other hand, we have spoken to many traditional farmers who say pretty much the same thing. A species of plant or animal, or fungus or bacterium for that matter, seems to have a place in the overall natural system. At one level the idea is pretty trivial – corals live in coral reefs, trees live in forests and so forth. But there is something more profound, recognized by both Darwin and many traditional farmers. There are particular conditions required for a species of crop to grow, and knowing what those conditions are enables the farmer to choose what crops to grow. There are particular conditions required for a particular pest to prosper, and knowing what those conditions are enables the farmer to avoid them. These conditions, in modern parlance, define the limits of the ecological niche of that pest or that crop.

The idea gained scientific legitimacy through the work of two field ecologists, and, not surprisingly, took on slightly different connotations for both of them. One of the acknowledged founders of the Western tradition in ecology was American ecologist Joseph Grinnell, who conceptualized the niche as a kind of physico-chemical envelope within which the species could exist. Taking a slightly different perspective, British ecologist Charles Elton had the idea that a species had some sort of a "role" to play in the ecosystem. Elton's conceptualization of the niche reflected more modern ideas such as "response and effect"[7] or niche construction.[8]

However, the modern niche framework must be attributed to Hutchinson, who formulated the idea as a "multidimensional hypervolume," a formulation that seems to have survived and become a standard of contemporary ecological literature.[9] Beginning with ideas akin to those of Grinnell, Hutchinson suggested we contemplate the limits beyond which a particular species is unable to persist. His formulation was perhaps a bit more physico-chemical than it needed to be, but for heuristic purposes it suffices to imagine a temperature range within which some particular species is able to survive. An arctic fish, for example, will survive over a range of average temperatures that is considerably lower than the range over which a tropical fish will survive. Likewise, an ocean fish will survive over a range of average salinity that is considerably higher than the range over which a freshwater fish will survive. The situation becomes interesting when the niche ranges of two species overlap, as suggested on the temperature axis in Figure 2.9, and on both temperature and salinity in Figure 2.9c. It should be emphasized that we illustrate the idea for two variables (temperature and salinity), but Hutchinson's idea extended to a large number of physico-chemical and biological axes such that, rather than an area representing the niche, a volume, indeed a volume in multiple dimensions (a hypervolume), came to represent this idea of the niche. So, the Hutchinsonian niche is a "multidimensional hypervolume." While we, as human beings, are unable to visualize anything beyond three dimensions, conceptually and mathematically there is nothing to restrict us in the number of dimensions to be considered.

Hutchinson effectively formalized what Grinnell had suggested. Species have certain environmental limits that they cannot transcend under normal circumstances. However, as Darwin and naturalists before him noted, and as is evident

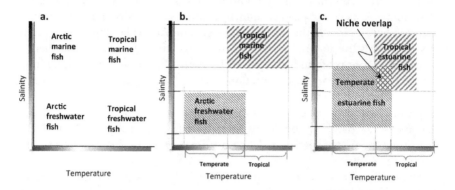

Figure 2.9 Illustration of Hutchinson's framework for the ecological niche. a. Basic limits on where four species of fish could live – two arctic fish that live in cold water, and two tropical fish that live in warm water, two marine fish that live in high salinity water, and two freshwater fish that live in low salinity water. b. The combination of the two environmental factors to describe the niches of two of the species, the arctic freshwater fish and the tropical marine fish. c. An example where the two species overlap in some parts of the niche space – note that the niches along the temperature gradient only already overlapped in the example in (b), but there was no overlap ultimately because there was no overlap along the salinity gradient, but here there is overlap on both gradients, since the fish are estuarine.

to every farmer, species that do the same thing, or similar things, do not coexist for long periods of time. That is the essence, for example, of plants called weeds. When rice or corn or wheat tries to occupy the same field with other grass species, unless the farmer intervenes with weed control, the crops usually lose out. This is akin to saying that the crops and weeds occupy similar niches (i.e., have similar requirements), at a fundamental level, and because they do, the weeds usually take over from the crops (although the opposite occurs also, but is effectively not noticed). This idea has been around for many years, intuitively understood by traditional farmers and codified in the science of ecology by the Russian physiologist Gause.[10] It basically says that if the fundamental niche of two species is the same, one of those two species will be eliminated through competition. It is usually known as the "competitive exclusion principle" or simply Gause's principle, and is certainly intuitively obvious, at least at its extreme.

However, Hutchinson elaborated the idea of the niche in a broader context and noted that the fundamental niche of a species is not where the species is actually found. Polar bears *could* live in the Amazon, but they don't. German cockroaches *could* live in the rain forests of Puerto Rico, but they don't. Morning glory vines and maize have almost the same basic requirements, but they do not live together (unless farmers cut back the vines repeatedly). Such observations make it clear that the fundamental niche of Hutchinson does not tell us where a species will be found. Rather, the ideas of Elton must enter the equation, which is to say the species interactions with one another must be taken into account, including any transformations they induce on other aspects of the environment. That is, the actual space "realized" by the species is a consequence of its fundamental niche modified by its interaction with other species and the environment. Thus, Hutchinson recognized these two ideas, the fundamental niche and the realized niche, and for the most part we still today use these basic ideas.

The whole idea is frequently conceptualized in terms of ecological competition, however it has much broader meaning. It seems reasonable to suggest that two species are more likely to compete more strongly to the extent that their fundamental niches "overlap." In modern ecological literature this is an idea that, while it has its detractors, effectively is the standard way of thinking about niches and competition. The degree to which two species' fundamental niches overlap is proportional to the intensity of the competition they exert on one another. Expanding a bit on our original example, we note the difference between fundamental niche and realized niche through competition between two estuarine fishes in Figure 2.10.

A further complication, clearly recognized by Hutchinson, but rarely emphasized, is the broad way in which Elton thought of organisms affecting not only one another but also the environment in which they live. It is a perspective that has long been recognized. For example, the idea is implied in Darwin's famous observation that plants have, as part of their niche (he never used the word itself), the bees that must pollinate them, and that bees tend to nest in abandoned mouse nests. But he went even further by noting:

> the number of mice is largely dependent . . . on the number of cats. . . .
> Hence, . . . the presence of a feline animal in large numbers in a district
> might determine, through the intervention first of mice and then of bees,
> the frequency of certain flowers in a district!"

(Darwin, 1859, Chapter 3)

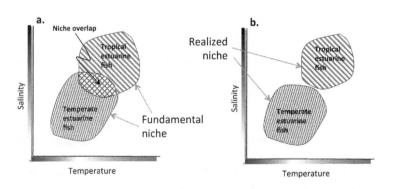

Figure 2.10 Diagrammatic representation of the transformation of fundamental to realized niche through the process of competition. The area of the space (temperature versus salinity) where the niches overlap is the area in which the two species cannot coexist, and thus one or the other will win in competition, basically leaving the two realized niches completely non-overlapping.

Clements, a major influence in US ecology in the early twentieth century, noted:

> By the term reaction is understood the effect which a plant or a community exerts upon its habitat . . . It is entirely distinct from the response of the plant or group, i.e., its adjustment and adaptation to the habitat. In short, the habitat causes the plant to function and grow, and the plant then reacts upon the habitat, changing one or more of its factors in decisive or appreciable degree.
>
> (Clements, 1928, p. 79)

Fifty years later, botanist John Harper stated:

> A plant may influence its neighbors by changing their environment. The changes may be by addition or subtraction. . . . There may also be indirect effects, not acting through resources or toxins, but affecting conditions such as temperature or wind velocity, encouraging or discouraging animals and so affecting predation, trampling, etc.
>
> (Harper, 1977, p. 354)

Some years later, in writing about intercrops, one of us wrote:

> [an organism] lives according to the dictates of its local environment, yet is an important participant in effecting change on that local environment.
>
> (Vandermeer, 1989, p. 10)

More recently a group of evolutionary biologists have taken the idea to a more controversial level in applying it to the basic dynamical structure of organic evolution, with the idea of "niche construction."[11] Whatever the outcome over the current debate about niche construction in evolutionary time, its importance in ecological time cannot be dismissed. As described in the most recent authoritative account (Wikipedia),

> **Niche construction** is the process in which an organism alters its own (or other species') environment, often but not always in a manner that increases its chances of survival.
>
> (retrieved August 16, 2016)

Finally, in what we regard to be the most elegant elaboration of the idea, Levins and Lewontin spoke of "constructivism" as a general phenomenon in biology, in which evolution, organic development, and ecology intersected in such a way as to create what might be called the dynamics of life. This viewpoint acknowledges the complexity of the issue. For example,

> if two species are . . . mutually changing the conditions of each other's existence, then the ensemble of species or of species and physical environment is an object with dynamic laws that can be expressed only in a space of appropriate dimensionality. The change of any one element can be followed

as a projection on a single dimension of the changes of the n-tuple, but this projection may show paradoxical features, including apparent lack of causality, while the entire ensemble changes in a perfectly regular manner.[12]

Although we acknowledge the idea as grandiose and perhaps visionary for the future of biology (historians will have to weigh in later), in the context of the present text, the dialectical interplay between "environment" and "organism" in ecological space and time is critically important. Despite its contemporary complexity in a wide-ranging literature, from the effect and response of Clements to the evolutionary consequences of the niche constructionists, it seems to us that Hutchinson had it well-summarized in his notion of the fundamental and realized niche. The fundamental niche exists within the limits imposed by the world, and the realized niche results from the organism's actions within and effects on that fundamental niche. But this process is dynamic. A predator's fundamental niche must include something for it to eat, its prey, but by eating the prey, the nature of that requirement is changed (lowered numbers of prey), to the point that sometimes the predator renders its own fundamental niche empty (eats all the prey, creating an environment for itself with no food). A legume plant growing in competition with a non-legume may modify the soil conditions such that its non-legume competitor may take advantage of the extra nitrogen supplied by the legume, effectively creating its own empty realized niche.

Viewed in this way, there is a kind of "chicken and egg" problem. Which came first, the niche or the organism? The answer, as Levins and Lewontin so eloquently note, is neither – organism and niche are dialectically constructed. The niche is "constructed" by the organism and the organism is "constructed" by the niche. This constructivist point of view is simultaneously confusing and clarifying. It is confusing in that strict cause-and-effect thinking, so important to scientists in the past, is challenged. To say that the realized niche emerges from the dialectical interaction between the organism and its fundamental niche, on the one hand sounds intuitively pleasing and makes sense of many observations one may make in nature. Yet, how to operationalize both fundamental and realized niche is less than crystal clear. We leave the problem at that.

Regardless of its unpleasant inner contradictory nature and philosophical complexity, we find it of use both in talking to people about ecology and in formulating more complex ideas about ecology. We also think it is a common mode of thinking among traditional farmers, who resist the simplistic cause-and-effect thinking of much formal natural science. It is a simplification, albeit a useful one, to begin with the fundamental niche and construct the realized niche. However, at its most analytical level, it is the organism/environment dialectic that jointly determines the nature of one another, in an endless feedback cycle.

Population dynamics

The very word dynamics sometimes confuses. It simply means change in the number of individuals (or amount of biomass) in a population. The numerical state of a population can be said to be 10 individuals, 100 individuals, 10 kilos of individuals, 25 nests of a species of ants, the number of plants infected with a disease, and any other number of possible measures of the size of a population. The dynamics of a population refer to the way in which the population state changes over time. So if we go from 100 individuals this year to 150 individuals in 2 years' time, the rate of change of the population is 25 individuals per year (150 −100 = 50, divided by 2 years) on the average. If we go from 50 plants infected with a disease this year to 20 plants infected 3 years from now, the rate of change of the disease population is −10 plants per year (negative rate of change). The word dynamics simply means change over time.

The pattern of change is very frequently the point of departure when trying to understand a population. A pest insect, for example, may become a pest only when it reaches some threshold abundance, and that threshold will be predictable if we know something about its population dynamics. Most often the dynamics are represented as a per capita rate (r), that is, how many individuals are produced in a single time unit by an average individual, or how many kilos produced from a single kilo and so forth. So, if we have a population of 10 individuals and each of those individuals produces two individuals per week, in one week we will have 20 individuals. If the rate is constant (a very important point) in the following week we will have 40 individuals and the week after that 80. So we can easily see the series 10 to 20 to 40 to 80 to 160 and so forth. But we could also imagine that on average a single individual will produce on average fewer than two new individuals per time unit, say 1.8 individuals on average (an average 10 individuals will produce 18 offspring in one time unit). Both of these situations are presented in graphical form in Figure 2.11.

The per capita rate of change is normally referred to as the intrinsic rate of natural increase. If it remains a constant value, no matter the initial size of the population, the pattern over time always has the basic form as shown in Figure 2.11. This particular pattern is called exponential growth and will always occur when the intrinsic rate is a constant (no change in its value as the size of the population changes). It is the basic form of population dynamics when per capita rate is constant.

There is a fundamental problem when we confront actual populations with this model. Suppose for example that we begin with a population of 100 bacteria that double every hour. A population of bacteria growing at such rate will generate the following sequence: 100, 200, 400, 800, 1600 and so forth. After three days there will be more bacterial cells than there are stars in the universe, a result of exponential growth. Most observers find this situation a bit unrealistic.

It is generally assumed that a population will eventually reach a point where the resources available are not enough for the population to keep growing. At that point, frequently referred to as the carrying capacity and symbolized by K, the actual growth rate of the population is zero, neither increasing nor decreasing. What we normally do is presume that there is a gradual approach to this point. That is, when the population is, say, halfway between a very small

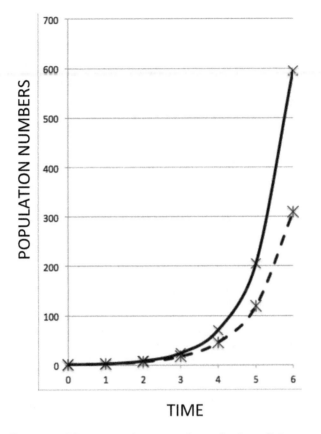

Figure 2.11 Illustration of the pattern of exponential growth of two distinct populations

value (near zero) and its carrying capacity, its growth rate is also about halfway between its maximum value (when the population itself is small) and zero (at the carrying capacity). A population behaving in this fashion is illustrated in Figure 2.12a, along with the comparative population that might be changing according to the exponential law (dashed curve). The pattern is referred to as logistic growth.

Growth can be either positive or negative. In Figure 2.12b we illustrate four different starting points for a population with the same basic rules (the same intrinsic rate of increase and the same carrying capacity). Note how the populations that are initiated above the carrying capacity decrease toward that carrying capacity, while those below the carrying capacity increase toward it. The carrying capacity (precisely 100 individuals in Figure 2.12) is thus called a "stable equilibrium point" for the population. This is a key concept for

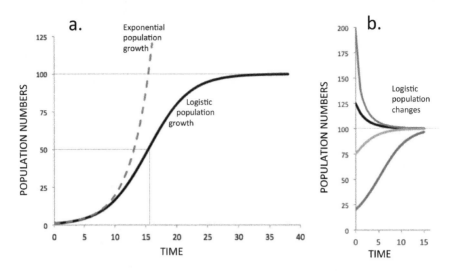

Figure 2.12 Illustration of logistic population changes. a. The form of logistic growth (solid
black line) compared to the exponential form (dashed line). b. Four populations
initiated at four different points, two of which are above the carrying capacity
and thus decline toward it, and two of which are below the carrying capacity and
thus increase toward it. The carrying capacity is an "equilibrium," in that once a
population arrives there it will remain in place in perpetuity. It is also a "stable"
equilibrium point, since any deviation (either above or below) will result in a
return to the point (an idea further explored in the next section).

understanding ecological complexity, as we present in considerable detail in
the next section.

Equilibrium, resilience, persistence

Ecologists sometimes imagine a "pristine" state, in which the "balance of nature"
is a reality. This viewpoint is largely incorrect. Modern ecology has come to
realize that, much like human political history, things change, sometimes very
rapidly, sometimes very slowly. In human history this phenomenon is sometimes
referred to as the *longue durée*, meaning the time during which contemporary
sociopolitical arrangements appear to us to be permanent (e.g., herding societies, hunter/gatherer societies, feudalism, capitalism). It is true that such times
exist, but it is also true that what seems permanent at one point in time, when
viewed from a historical perspective, turns out to be not so permanent. You may
be reading this in electronic form right now. That statement would have made
little sense to your great grandfather, and whatever we say in this book will likely
seem quaint and naïve to our great grandchildren. Ecological systems are similar.
They tend toward stable equilibrium states sometimes, they change sometimes,
and they persist in a *longue durée*, even though they continuously change in their
specifics and sometimes rupture into a new regime entirely.

But modern ecology has long rejected the idea that there really is such a thing as a "balance of nature" or that nature is in "equilibrium" in some idealized way. Yes, persistence of what we see in nature is sometimes visible; for example, a coral reef or tropical rain forest appears generally the same from year to year. But dramatic change is sometimes also visible: a savannah or prairie is subject to natural fires periodically, a hurricane seems to do major damage to both coral reefs and rain forests, and so forth. Part of the underlying theoretical structure for thinking about such things has to do with the formalities of concepts such as equilibrium and resilience. Equilibrium means that the forces affecting a system balance one another such that the system stays in its current state forever. This could be, for example, a boulder at the bottom of a valley or a boulder at the top of a hill. Being in the valley or top of the hill, however, are very different things from the boulder's "point of view." At the top of the hill a small shove might cause the boulder to tumble to the bottom of the valley, but once nestled in the valley, even a large shove will result in a return to the bottom of the valley. While both situations are "in equilibrium" (the forces, prior to the shove, balance one another out) the boulder at the bottom of the hill is at a "stable" equilibrium, while the boulder at the top of the hill is at an "unstable" equilibrium.

This general idea is frequently illustrated with a physical model.[13] For example, in many tropical agroecosystems trees are an important part of the system (called agroforestry). If the trees are legumes, they may very well house nitrogen-fixing bacteria and contribute to the nitrogen available in the soil, making the understory crops grow faster. However, the trees also create shade, thus reducing the rate of photosynthesis in the crops (Figure 2.13a). If the system is located at the legume tree density indicated by the dashed vertical line in Figure 2.13a, a move to reduce the trees by the farmer will lead to reduced absorption of nitrogen by the crop and increased photosynthesis (see the horizontal arrow in the left-hand graph of Figure 2.13a). However, if the relationship between yield and photosynthesis and yield and nitrogen absorption is arranged as shown, the reduction in yield due to nitrogen absorption is much greater than the increase due to reduced photosynthesis, meaning that the overall yield of the crop (the sum of the two curves in Figure 2.13a) will be reduced. Experiencing that the yield is reduced, the farmer logically will plant more trees. If the deviation from the vertical dashed line had been to the right, a similar yield reduction would have occurred for the reversed reason (high drop in yield increase due to photosynthesis and low yield increase due to nitrogen absorption), causing the farmer to reduce the density of the trees.

If the form of the relationship between photosynthesis and nitrogen absorption were different, as in the case of the right-hand graph of Figure 2.13a, we can see that similar deviations from the dashed line would lead to quite different results. Beginning at the equilibrium point (the dashed vertical line in Figure 2.13a), reducing the tree density would cause an increase in yield due to a high increase in yield due to photosynthesis and a low decrease in yield due to a decline in nitrogen availability, meaning that the farmer, if responding only to yield, would likely continue eliminating trees from the system. Alternatively,

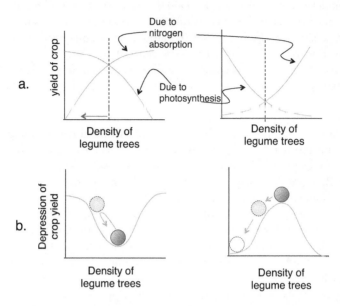

Figure 2.13 Stylized diagram of crop yield as a function of an overstory of legume shade
     trees. a. The yield of the crop as a function of the density of legume trees.
     b. Transformed version of (a), wherein the depression of the crop yield is plot-
     ted as a function of the density of legume trees, suggesting a physical model of
     a curved surface on which a ball moves. The left-hand side illustrates a stable
     equilibrium point; the right-hand side illustrates an unstable equilibrium point.

beginning at the equilibrium point in the figure on the right-hand side of
Figure 2.13a, increasing tree density would cause a much lower yield decrease
due to reduced photosynthesis than the yield increase due to increased nitrogen
availability, meaning that the farmer would likely continue planting trees in the
system.

In both graphs of Figure 2.13a the system is at equilibrium (the dashed vertical line), which simply means that there is a balance of forces that keeps it at
its present state. However, the left-hand graph represents a "stable" equilibrium
point while the right-hand graph represents an "unstable" equilibrium point.
The concept of stable versus unstable equilibrium is frequently represented with
a physical model of a surface on which a ball rolls (similar to the boulder in a valley versus a mountain top earlier in this section). We can imagine such a surface
as a simple transformation of the photosynthesis/nitrogen absorption curves of
Figure 2.13a. In Figure 2.13b we represent those same data as the "depression
of crop yield." We can then imagine that the resulting curves (Figure 2.13b) are
surfaces on which the physical force of gravity applies. The dynamics of the
system thus emerge if we try and balance a ball on that surface (Figure 2.13b).
In the case of the left-hand graph (Figure 2.13b), if we push the ball up on the
side of the center well, it immediately comes back (as it did when the farmer

decided to reduce the number of shade trees). Alternatively, in the case illustrated in Figure 2.13b on the right-hand side, if a ball balanced on the top of the curve is pushed only slightly, it continues on a trajectory downward. The distinction between the equilibrium on the left-hand side and the equilibrium on the right-hand side is the distinction between a stable versus an unstable equilibrium point.

In addition to the question of the stability of an equilibrium point, several other key concepts have populated the ecological literature in recent years. Two that we feel can be important to understanding the complex ecology of agroecosystems are "resilience" and "persistence." Both are easy to conceptualize if we employ the physical modeling framework of the ball on a surface, as illustrated in Figure 2.14. Comparing high resilience to low resilience (Figure 2.14), the important feature is that if perturbed, high resilience results in a return to the same stable equilibrium point even if the perturbation is large. In contrast, low resilience means that the ball can be shifted to a different equilibrium point if the perturbation of it is large enough. Persistence simply means that, even in a locally low resilient situation it will be difficult to predict which precise equilibrium point will actually contain the ball. Overall the system will reliably persist in the larger well.

The simple ball and well metaphor of Figure 2.14 is useful for introducing the ideas of equilibrium, stability, resilience and persistence. However, by forcing the explanation to appear on a flat page in this limited two-dimensional form, something very important is left out: the idea of oscillations. Suppose, for example, that we have a swinging pendulum. It swings from left to right, then from right to left in a slowly decreasing rhythm that eventually winds up in a straight-down position, no longer oscillating (as in Figure 2.15a). We could represent its behavior as a ball and well diagram, wherein the weight of the pendulum swings back and forth to the sides of the well (as in Figure 2.15b), but if we do so we sort of lose one of the main features of the pendulum. It doesn't really jump back and forth, it oscillates. What is usually done when analyzing the pendulum is to consider two variables: (1) the angle the pendulum makes with the straight-down position, and (2) the velocity at which it moves. Then, rather than having the simple ball and well model as in Figure 2.14 and 2.15b, where x is the variable of interest and y is some measure of the potential of the system to move, we have two variables of interest (angle and velocity) and the "potential"

High resilience   Low resilience   Persistence

Figure 2.14 Illustrations of the key concepts of resilience and persistence with a physical model of ball on a surface

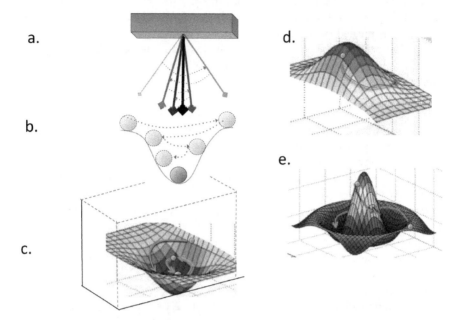

Figure 2.15 Representation of the basic oscillatory process as "response" surfaces

is a third dimension in the system (low potential means a slow movement toward the equilibrium, while high potential means a rapid movement toward the equilibrium). Plotting the pendulum in this way results in a three-dimensional well, again with a metaphorical ball that moves on its surface. As pictured in Figure 2.15c, the expectation is that the ball oscillates from the edge of the well to the bottom of the well, and the previous picture (in Figure 2.15b) is a cross section of that surface. Of course not much is different qualitatively from the two dimensional picture of Figure 2.14, but with this more complete picture another, extremely important, dynamic behavior can be qualitatively visualized.

To see this other behavior we first add the idea that an equilibrium may be oscillatory in an unstable way in addition to the stable way of Figure 2.15c. We picture the relevant surface in Figure 2.15d. As you can see, now we have, in three dimensions, the equivalent of the two-dimensional illustration on the right-hand side of Figure 2.13b. Combining Figure 2.15c and 2.15d, we can imagine the relatively complicated surface area pictured in Figure 2.15e. Here if we begin at either of the unstable parts of the diagram (either in the peak or on the edge of the rim, the system will oscillate, but there is no single point to which the oscillations will tend. That is to say, the system, in the limit (after a very long time) will be at a particular cycle, the cycle that is at the bottom of the diagram. This behavior is called a "stable limit cycle" and it characterizes many ecological phenomena. More importantly, it is a fundamental organizing feature of several emerging issues of ecological complexity.

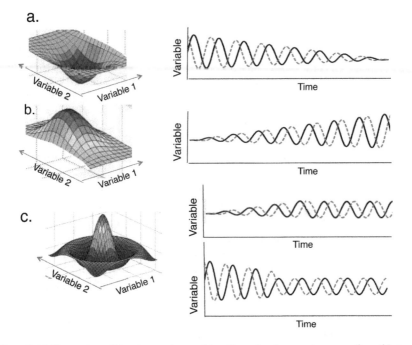

Figure 2.16 Illustration of the time series resulting from the three main types of equilibrium
situations. Note that there are two resultant time series. The top time series results
from one particular starting, while the bottom series results from a distinct starting, while the bottom series results from a distinct starting point.

If the previous description was not crystal clear, perhaps an alternative presentation of the patterns will help. In Figure 2.16 we present each of the three
qualitatively distinct patterns (stable oscillations, unstable oscillations, and limit
cycle) along with a graph over time (a time series) to illustrate the generalized
pattern in all three cases.

Basic trophic dynamics

We have already indirectly discussed the idea of trophic dynamics. When decomposing bacteria consume a piece of dead organic material, the energy and material
of that dead organic material is transferred to the bacteria. Transferring energy and
materials from one organism to another emerged as a key concept in ecology with
the publication of an extremely important article by ecologist Raymond Lindeman
in 1942 . Some ecologists even view this point in time as the birth of a division in
ecology between ecosystems ecology and population/community ecology.[14] As
we mentioned before, we think this division has not been all that productive for
the development of the science in general (the book *The Phytochemical Landscape*
by Mark Hunter is an excellent antidote to this division).

The key dynamical features of trophic dynamics can be most easily visualized when thinking of the basic process of predator and prey (at a most generalized level they are identical to herbivore/plant, disease/host, parasite/host, saprophyte/organic matter, and probably others – any situation where one biological unit transfers energy to another). A simple graphical analysis illustrates the most underlying significance of trophic dynamics. Consider the common example of aphids and their predators, frequently lady beetles (ignore for the moment the transfer of energy and materials from the plant to the aphid). There are four situations that matter: (1) a field that contains large numbers of aphids and very few beetles; (2) a field that contains large numbers of beetles and very few aphids; (3) a field that contains large numbers of aphids and large numbers of beetles; and (4) a field that contains very few aphids and very few beetles. If there is a large number of aphids, the beetles will have a feast and probably have a high reproductive rate (and low death rate), but if there are very few aphids the beetles will begin dying of starvation. From the other point of view, if there are a large number of beetles around, the aphids will get eaten rapidly and their population will decline, but if there are very few beetles around, the aphids will prosper without the specter of getting eaten. These observations can be summarized on a graph of the population density of prey (aphids) versus predators (beetles), as in Figure 2.17.

The representation of a system like this as in Figure 2.17c is called "phase plane analysis," since the two "phases" of the system are the two axes of the represented space (the "plane"). In contrast, the representation as in Figure 2.17d is called a "time series" and represents precisely the same data but in a different form: population numbers over time. In our experience, when first encountered, in the time series representation it seems wholly obvious that the system is oscillating, whereas in the phase plane presentation the loops that are generated can cause some initial confusion and obscure the main point that the system is oscillatory. Yet, in ecology it has become commonplace to analyze systems in the phase plane since it is frequently quite easy to arrive at qualitative conclusions from simple manipulations of the expected arrows (formally referred to as "vectors," the collection of which in any phase plane is called the "vector field"). The conclusion in this case, a completely necessary consequence of the underlying qualitative nature of one thing, the predator, eating another thing, the prey, is that such systems will have a tendency to oscillate. In the real world we have to take all sorts of complications into account, but underneath it all, oscillations are to be expected.

It will undoubtedly have occurred to some readers that there must be some value that separates the increasing from the decreasing vectors (the arrows in Figure 2.17a, b, and c). That value we will call a "zero growth isocline" (sorry, we couldn't think of a simpler term, so let's just go with the formal terminology), sometimes referred to with the simple shorthand "isocline." As we mentioned before when describing population dynamics, the point at which a population ceases to grow is called the carrying capacity. Therefore, the zero growth isocline for each species separately represents the carrying capacity of one population. In

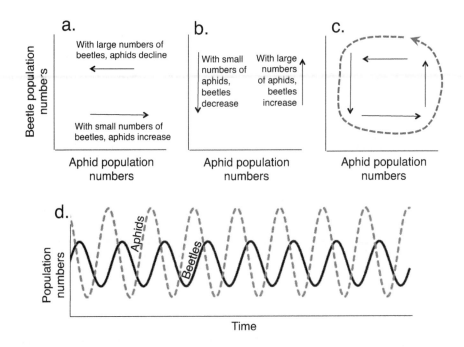

Figure 2.17 General form of the predator/prey relationship (the fundamental trophic interaction). a. What is expected at either high or low numbers of the predator
(beetle). b. What is expected at either high or low numbers of the prey (aphid).
c. Combining (a) and (b) in a single graph. Note that by simply following
the arrows we can generalize that the system will form a sort of loop, which
is another way of saying that the system will oscillate. d. Representation of the
picture in (c) as a time series, where the oscillatory nature of the two species
can be easily discerned.

Figure 2.18 we illustrate the idea for the beetle/aphid trophic dynamic example,
effectively repeating the previous explanation (Figure 2.17) but with the use of
isoclines (dashed black lines in Figure 2.18). Not surprisingly, we conclude that
the system will oscillate.

   In this section we have noted the basic trophic rule that a consumer eats a
resource. We hasten to note that there is no ecosystem in the world (as far as we
know) where such a relationship exists in a vacuum. Indeed, we find it almost
impossible to even imagine a species that does not have, at a bare minimum,
both a predator and a disease (remember, a disease is caused by an organism
which sequesters energy from its host, which is to say, just a slightly different
form of a trophic connection). So our idealistic version of a single consumer
and a single resource should be thought of as a very simple base-line beginning for understanding how energy flows through, and materials cycle around,
ecosystems.

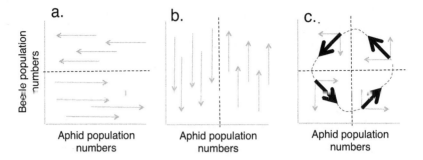

Figure 2.18 Exploring oscillatory dynamics with isoclines. The vertical and horizontal dashed lines are the isoclines of the system, and the arrows are members of the vector field illustrating the main flow of the system for (a) the aphids and (b) the beetles. In (c) we show the combined effect of the two dynamical rules (one for predators, one for prey) where the bold black vectors represent the sum of the horizontal and vertical vectors and indicate the general flow of the system (the dashed trajectory is one oscillatory cycle).

Feedback in dynamic systems

The isoclines in Figure 2.18 are what might be called neutral isoclines. They assume that there is a critical number of predators that determine whether the prey population will increase or decrease. This neutrality assumption presumes that the individuals in the prey population do not interfere with one another, an assumption that certainly cannot be true in general. In an overly stocked pasture, for example, we all know that the pasture becomes overgrazed and the livestock suffer. And as soil bacteria absorb nitrate from a static soil, their population eventually becomes large enough that there is no more nitrate left, disabling any new reproduction of the population. Each individual in a population consumes resources at a rate commensurate with its needs and abilities, and a single cow in a small pasture does just fine, whereas if it must share that small pasture with, say, 100 other cows, forage will become limiting. If we think of the population as growing at some particular rate, as the resources that it needs become used up, there is effectively a transfer of information from that population to the rate of growth, a feedback in the system. So if the size of a population (i.e., number of individuals) at some time in the future depends on the population size now plus the growth rate of that population, that is saying that the growth rate in numbers is a function of the numbers themselves. There is a feedback from the variable of concern (i.e., the population density) and the rate of population growth. Such a feedback, in the case of a growing population, results in the cows (or bacterial cells) competing with one another for what has become a limiting resource. This form of competition is referred to as "intraspecific" competition (to distinguish it from competition between two different species, "interspecific" competition), and is probably the most significant form of feedback in ecology, and is the basis of the carrying capacity formulation introduced earlier. But of most importance is the fact that such a feedback does not just "regulate" some process, as when we go from the exponential population to the logistic population, it frequently transforms the qualitative nature of the process itself. A simple trophic dynamic system is a case in point, as we now explain.

Just as predators exert a negative effect on the rate of population change of their prey, it is reasonable to presume that the prey individuals will sometimes exert a negative effect on themselves. If we were to repeat the diagram of Figure 2.18a but where the population density of the prey were substantially higher, we might think of the upper (right-hand side) densities of the prey as contributing to the construction of the critical density of predators. So, for example, begin with a population of 100 aphids in a small bean field. It might take a population size of 100 or more beetles to cause the aphid population to begin to decline and eventually to lead to local extinction of the aphids. But if there were 1,000 aphids in the same bean field, the condition of the aphid population might be very near to a separate tipping point simply because of the competition among those aphids (the in*tra*specific competition). Consequently, it might take a population size of only 50 beetles to tip the balance and cause the aphid population to begin its decline. In effect, at higher population densities of aphids, the critical predator density will generally be lower, as illustrated in Figure 2.19a. It takes little further argument to generalize, making the isocline a declining function of the prey density, as in Figure 2.19b. Note the addition of the parameter K.

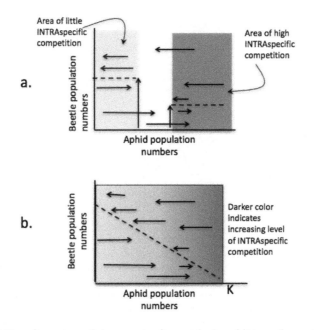

Figure 2.19 Transformation of the prey isocline with the addition of prey intraspecific competition

A moment's reflection reveals K to be the aphid population density for which the aphid population will neither increase nor decrease. This is a key parameter in population ecology, commonly referred to as the "carrying capacity" of the environment, as we discussed in a previous section (population dynamics).

Trophic dynamics change dramatically from this addition. Although the important conclusion that trophic systems will oscillate remains true, those oscillations change in their character due to this addition. To see this, we alter the original neutral prey isocline with the new declining prey isocline in Figure 2.20. Concentrate on the shaded quadrants in Figure 2.20a. Note that the vectors point up and to the left in the upper right quadrant and down and to the right in the lower left quadrant. In Figure 2.20b we changed the prey isocline to represent intraspecific competition (as in Figure 2.19b). Note that the area of the phase space in which those original arrows point up and to the left or down and to the right have expanded, such that the original trajectory (the dotted oscillations) tends to be pushed toward the equilibrium point (where the two isoclines cross). This results in oscillations that gradually approach the equilibrium point, still oscillating, but with lower and lower amplitude, as illustrated in Figure 2.20c and d.

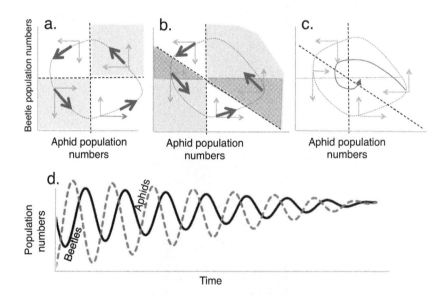

Figure 2.20 Adding intraspecific competition to the basic oscillatory pattern of trophic dynamics. a. The original generation of oscillatory behavior. b. Changing the prey isocline to reflect the reality of intraspecific competition. Note the expansion of two of the areas of the phase space to change how much of the space includes vectors that effectively cause the oscillations to move toward the equilibrium point. c. Illustration of how the ultimate consequence is that the oscillations, while still there, will gradually spiral in toward the theoretical equilibrium point. d. a time series illustrating the same dynamics as in (c).

Recall our discussion of density dependence in the prey. When the prey is very abundant, the critical density at which the predator will cause the prey to decline is amplified by the intraspecific competition imposed by the large number of prey individuals, as in the right-hand side of the graph in Figure 2.21a (also recall Figure 2.19). But we can also ask what happens at very low densities of the prey (the left-hand side of the graph in Figure 2.21a). If the prey is extremely close to zero (one individual), the critical number of predators required to cause that prey item to go extinct is probably itself very low, but, at any rate, is the critical value we spoke of earlier when talking about the original prey isocline (e.g., as in Figure 2.19a). But adding a bit of complexity, if it takes, for example, 10 predator individuals to drive a population of one prey extinct, what might it take to drive a population of 10 prey extinct? Probably more than 10 predator individuals. That is, at very low densities of prey, the critical value of the predator tends to increase as the prey density increases. So we have the situation, as illustrated in Figure 2.21a, that at low prey densities the prey isocline tends to increase (have a positive slope), while at high prey densities the prey isocline tends to decrease (have a negative slope). It then becomes obvious that the prey isocline has to have something like a humped shape to it, as illustrated in Figure 2.21b.

On the other hand, the position of the predator isocline is inversely proportional to the feeding rate of the predator. So, for example, if that feeding rate is very small, the predator isocline will fall to the right of the carrying capacity and it is easy to see how the predator cannot persist (predator goes extinct even if the prey is at its own carrying capacity). However, if the predator isocline is below the value of K but to the right of the peak of the hump of the prey isocline, the behavior of the system is a stable focal point (stable oscillatory point) as described earlier (Figure 2.20). However, if the predator isocline is to the left of the peak of the hump of the prey isocline, the oscillations are unstable, with their amplitudes forever expanding, as in Figure 2.21c and 2.21d. For most mathematical models that describe this particular sort of behavior, those ever-expanding oscillations actually turn into a limit cycle (as in Figure 2.16c), but the behavior of the system very near to the equilibrium point (where the two isoclines intersect) is unstable (ever-expanding oscillations).

A particularly important generalization for agroecology has to do with the idea of autonomous biological control. As many farmers and agroecologists have noted in the past, a farm that is well-managed will tend to be self-regulating with regard to the issue of pests, and the farmer or manager's goal is to seek management strategies that promote that sort of autonomous control.[15] But a glance at the graph in Figure 2.21c suggests that there could be a problem here. If the position of the predator isocline is proportional to the inverse of the feeding rate of the predator, it makes sense for the farmer to try and make that feeding rate more efficient (i.e., increase it). With whatever tools are available, then, the farmer or manager may increase the feeding rate, which moves the predator isocline to the left (see Figure 2.21b and 2.21c). Thus, ironically, in attempting to increase the efficiency of biological control, the farmer may create a more

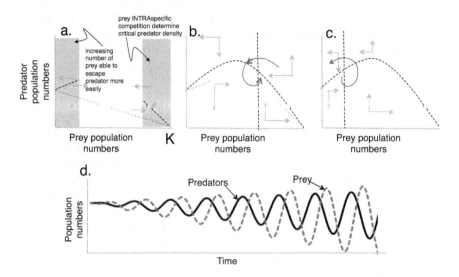

Figure 2.21 Effects of relaxing the assumption of simple density dependence. a. The changing response to the increasing number of prey able to escape more easily (at low
prey densities, but as the prey increase), versus the increasing tendency of prey
to control their own population through intraspecific completion (at high prey
densities). b. Resulting hump-shaped predator isocline and the dynamic behavior
of the system with a predator isocline to the left of the K, but to the right of the
peak of the prey isocline – a stable oscillatory situation. c. Same as (b), but with
the predator isocline located to the left of the hump of the prey isocline – an
unstable oscillatory situation. d. Time series illustrating the unstable oscillations.

volatile situation, in the extreme actually causing the extinction of the control
agent. This irony is known as the paradox of biological control.

The preceding result is important in that it illustrates the key role of feedback, in this case intraspecific competition, in changing the qualitative picture
of trophic dynamics. Yet one particularly important aspect of trophic dynamics occurs when more than one consumer is involved in the consumption
of a single resource. While the dynamic oscillations associated with a single
consumer and a single resource (as presented in Figures 2.17–2.21) are certainly important, things change dramatically when we have two consumers
in competition for the same (or overlapping) resources. For the most part, in
elementary considerations we do not consider the entire three-dimensional
system (resource plus two consumers), but rather suppose that the resource is
fixed or that it approaches its own equilibrium point extremely fast compared
to the rates of change of the consumers. Thus, each consumer, when isolated,
will approach some fixed carrying capacity, and we examine the dynamic
deviations from those carrying capacities for the whole phase space created by
plotting the consumer populations against one another. This is the fundamental
subject of INTERspecific competition.

Much of agroecology is concerned with interspecific competition. When weeds compete with crops, when different biological control elements eat the same pests, when fungi compete with bacteria in the decomposition process, when different strains of nitrifying bacteria use the same resource of inorganic nitrogen, all of these examples are of INTERspecific competition. The basic rules of interspecific competition can be deduced in a similar way as trophic dynamics, beginning with the initial casting in phase space, this time with species 1 and species 2 (Figure 2.22a and 2.22b). We begin with the axes of the phase space with the density-independent (without competition) dynamic relationship imposed. Thus, the vectors in Figure 2.22a represent the behavior of each of the populations in the absence of the other (density independent), in which case each of the populations will approach its carrying capacity. In Figure 2.22b we have added a couple of additional vectors, suggesting that even in competition with a very small number of the other population, an approach to the carrying capacity will be at least the approximate result. But what if the second species of aphid had a large, say at approximately 50% of its own carrying capacity, population density? The "effective" carrying capacity of the first aphid then would be determined not only by the INTRAspecific competition it experiences from its conspecifics, but also the INTERspecific competition it experiences from the other aphid species. Thus, we would expect it to equilibrate at some value below its own carrying capacity, as illustrated in Figure 2.22c.

If we keep applying this logic to different values of the population density of aphid 2, it is clear that we will trace out an isocline, a line (focus of critical values of aphid 2) that separates those areas of space in which the population of aphid 1 increases from those areas in which it decreases, as shown in Figure 2.23a. A similar exercise of changing population density of aphid 1 and observing how the isocline (point where aphid 2 neither increases nor decreases) emerges is obvious (see Figure 2.23b).

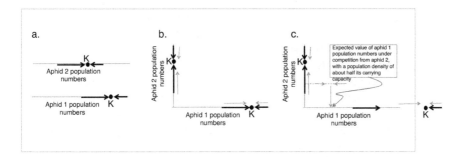

Figure 2.22 Basic population dynamics of two species. a. Population dynamics of each along the gradient of population densities, where each of them approaches its carrying capacity. b. Same as (a), but with the second species (aphid 2) rotated so as to envision the population growth in phase space. c. Same as (b), but with a position of the population of aphid 1 under competition pressure from aphid 2 illustrated on the *x*-axis.

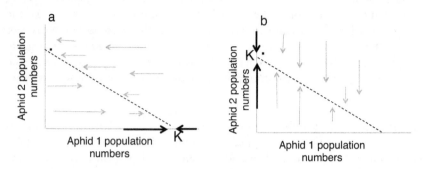

Figure 2.23 Extension of Figure 2.22c, in which the expected response of (a) aphid 1 and (b) aphid 2 are shown independently, but with all densities of the competing species illustrated (with the horizontal or vertical vectors)

Recalling Figure 2.18 when we combined the isoclines of the lady beetles and the aphids, we now do the same thing with the two aphids in interspecific competition. Here there are several interesting consequences. First, as can be seen in Figure 2.24, there are, for all practical purposes, only four ways that the two lines may be placed with reference to one another on the graph. This simple fact results in the very general conclusion that there are four qualitatively distinct outcomes of interspecific competition: (1) a stable equilibrium is established (Figure 2.24a), (2) species 2 always wins and excludes species 1 (Figure 2.24b), (3) species 1 always wins and excludes species 2 (Figure 2.24c), and (4) either species 1 or species 2 wins, depending on the starting point (Figure 2.24d). The last case is extremely important in ecology in general because it represents a very special type of equilibrium point.

The indeterminate case (Figure 2.24d) is frequently referred to as a saddle point, since a marble falling on a saddle will more or less behave in the predicted fashion (Figure 2.25). If it is balanced exactly at the equilibrium point, it will stay there forever. But with the slightest push in one direction or another it will fall off of one or the other side of the saddle (much like the unstable situation in Figure 2.13b). But the other important component of the dynamics is that if the marble begins at one of the higher points of the saddle (the darker, denser upper parts of Figure 2.25), it will begin by approaching the equilibrium point, but rapidly deviate to fall off one or the other side of the saddle, and the side to which it falls will depend on the initial position of the marble (if slightly to the right of the line connecting the two upper points on the saddle) it will fall to the right side; if slightly to the left of that line, it will fall to the left).

In the context of interspecific competition, the saddle point is the equilibrium point of one of the qualitative outcomes (Figure 2.24b). But saddle points emerge in many applications in ecology and sometimes are extremely important in guiding the intuition about the behavior of a system. For example, imagine that the exit from the response surface in Figure 2.25 connects with the part of

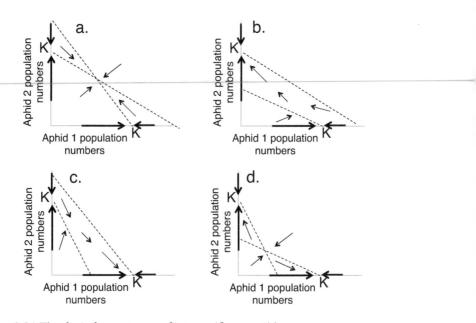

Figure 2.24 The classic four outcomes of interspecific competition

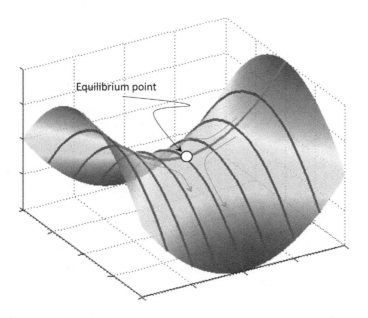

Figure 2.25 Response surface of a saddle point

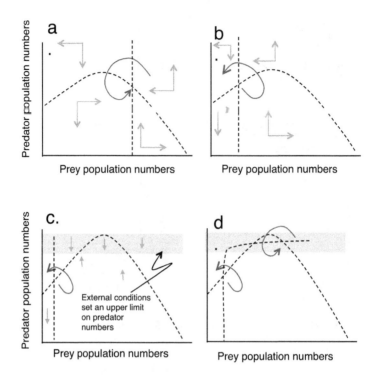

Figure 2.26 The logic of alternative possible equilibrium points in a consumer/resource
(predator/prey) situation. a. When the predator isocline (vertical dashed line)
crosses the prey isocline (dashed curve) to the right of the peak of the hump, the
outcome of the interaction is one of a stable focal point (or damped oscillations).
b. When the predator isocline crosses the prey isocline to the left of the peak
of the hump, the outcome of the interaction is one of an unstable focal point
(ever-expanding oscillations). c. Illustration of how external conditions may set a
limit on the predator numbers, meaning that if the oscillations produce a predator
population that is too large, those external conditions bring it back down, effectively creating a limit cycle. d. When the predator isocline is changed to account
for an ultimate limiting force on the predator (curves to the right at a very large
number of predators), it is evident that two stable equilibria are possible, one a
stable focal point (damped oscillation) and the other a limit cycle.

another saddle point in such a way that it moves in the direction of the equilibrium point of that saddle, but that second saddle has its output connected to the
input of the first saddle. This sort of arrangement emerges in many ecological
applications (sometimes in unexpected ways) and can represent the main structural element for the system as a whole (it is formally called a heteroclinic cycle).

Within the underlying theory of competition there is one set of conditions for
which alternative states are possible (Figure 2.24), where the state that is eventually approached depends on where the system is initiated. Species 1 may exclude

species 2, or species 2 may exclude species 1, depending on which species gets the initial advantage. A similar situation may emerge from a consumer/resource arrangement, as illustrated in Figure 2.26.

Notes

1 Some useful texts are Ricklefs and Relyea, 2013; Gotelli, 2008; Begon, Howarth and Townsend, 2014; Krebs, 2008; Vandermeer and Goldberg, 2013.

2 Biologists are in general agreement that there are three general life forms: eukaryotes, bacteria, and archaea. The eukaryotes include animals, plants and fungi and are most closely related to the archaea. The archaea were formerly thought to be just special forms of bacteria (having frequently been found in extreme environments they were sometimes referred to as "extremophiles"), but detailed analysis revealed that they are actually more closely related to eukaryotes than to bacteria.

3 Wittman, 2009.

4 Schmidt et al., 2012.

5 Dead zones are hypoxic (low-oxygen) areas generated by eutrophication (increases in nutrients) that lead to algal blooms and rapid consumption of oxygen, causing much of the natural aquatic/marine system to die.

6 For an introduction to this conceptually difficult controversy, see Whittaker et al., 1973; Chase and Leibold, 2003.

7 Goldberg and Landa, 1991.

8 Odling-Smee et al., 2003; Vandermeer, 2004b.

9 Hutchinson, 1957; Leibold, 1995.

10 Gausean niche concept and competition was first discussed in formal detail by Robert May in 1974, although its qualitative details were clearly articulated by Hutchinson in 1957.

11 Niche construction is one element of what some evolutionary biologists have come to refer to the "new synthesis." See for example Laland et al., 2015.

12 Levins and Lewontin, 2015, p. 134.

13 As noted by Scheffer, this idea is somewhat misleading in the end, but is useful to illustrate the basic principles; see Scheffer, 2009.

14 Ecosystems ecology is concerned principally with the quantities and rates at which tropic dynamics occur, population/community ecology is concerned more with the organisms involved and how they interact.

15 Lewis et al., 1997.

3 The Turing mechanism and geometric pattern

On a casual visit to a shaded cacao farm in northeastern Brazil, we encountered a farmer who tied strings from some of the shade trees to the adjacent cacao plants, effectively connecting the surrounding cacao plants with a single shade tree. When questioned as to his reason for doing so, he told us that nesting in the shade tree was a kind of ant that was very aggressive in controlling several kinds of pests that attacked his cacao plants, and that the ants would use the strings to gain access to the cacao plants. Not surprisingly, this reminded us of the common practice in ancient China of connecting citrus trees with bamboo sticks so that ants could travel from the tree that held their nest, to surrounding trees in the orchards, attacking all sorts of pests of the citrus tree.

This conversation with the Brazilian cacao farmer was especially important for our research program since we had recently discovered that the same genus of ant, *Azteca*, was a common presence in the coffee farms we were studying in Mexico, having its nests in about 3% of the shade trees on the farm. But in our case the situation was a bit more complicated since this ant is mutually associated with a pest species, the green coffee scale. Because even casual observations could easily convince any farmer (or us) that the ant was effectively a pest of coffee – protecting a well-known pest species and reflecting the old adage that a friend of my enemy is my enemy – the ant was generally regarded as a problem.

However, nature is, as so frequently asserted, more complicated than one's first impression. It turns out, as we explain in great detail in a later chapter, that the nests of these ants are grouped together in rather tight clusters, with nearby nests frequently within 4 or 5 m of one another, and vast sections of the farm with no nests at all. Any coffee bushes within one of these clusters will be generally tended by ants and will have high densities of the scale insects. But that is only 3% of the farm. In the rest of the farm, perhaps surprisingly, it is difficult to find the scale insects at all, due to a kind of beetle that eats them voraciously. The beetle adults have a remarkable ability to find solitary scales that are scattered throughout the farm. Yet, this extraordinary ability to find its prey even when the prey is extremely rare is not shared by the juvenile stage of the beetle. The beetle larvae are unable to fly and are effectively restricted to a single coffee bush. That is, although adult beetles fly and can locate their food over long distances, larval beetles crawl along leaves slowly, unable to travel even to nearby bushes. So if a larva finds itself on a coffee bush that contains only one or two scale insects (this species of beetle eats only scale insects), it will die of hunger, but if foraging among the large numbers of scale insects living under protection of the ants, it will thrive. The beetle larva is able to mingle among the aggressive ants because it is covered with waxy filaments that protect it against ant attacks. Thus, the clusters of ant nests provide the dense local concentrations of scale insects that allow the beetle larvae to grow, become an adult and fly around the farm eating all the scale insects on the rest of the farm. Thus, with only 3% of the farm under "ant management," 97% of the farm is provided with protection against this potentially important pest.[1]

It is the clustered nature of the ants that is critical to the provisioning of the ecosystem service of pest control. So, understanding the nature of that clustering becomes important for the efficient management of the system with regard to the control of the scale insect. Yet, more generally, the search for that type of understanding has deep roots in the history of ecology. A major goal of eighteenth- and nineteenth-century European explorations was the detailed mapping of natural patterns, including, for obvious economic purposes, those geological features that may have had some relationship to the minerals found under the ground.[2] Similarly, vegetation maps revealed zones of particular plant associations that, in turn, were assumed to result from basic soil and climatic conditions. Various schemes were devised to provide order to such vegetation types, most of which were ultimately based on a combination of temperature and rainfall, although soil factors loomed large in anticipation of agricultural applications. Such surveys, and some even today, assume that spatial patterns are determined by physico-chemical forces, effectively ignoring the possibility that biological interactions might be involved. Two more recent narratives help clarify some aspects of this perhaps anachronistic vision.

First, ecologists, farmers and pretty much anyone who has made any observations about nature are aware of the fact that biological organisms are not simply scattered across space randomly. Whether a snapshot of a population of elephants or a long-term study of the spatial distribution of corals on a reef, these observations indicate that almost all organisms are distributed in a non-random fashion.[3] Most evident is the frequently cited patchwork of vegetation in many arid and semi-arid parts of the world (Figure 3.1a). For many of these patterns, there are underlying biological processes that can be tied to their generation (Figure 3.1b), while others are clearly the direct result of underlying physico-chemical factors.

Second, and somewhat more subtle, there has long been an appreciation that particular species can be identified with an ecological niche, variously described as "an organism's place in nature" or the "role of an organism" or a similar intuitive device. Usually identified with either the British ecologist Charles Elton or the American ecologist Joseph Grinnell, the ecological niche has been a centerpiece of ecological thinking since the early twentieth century.[4] But it was not until 1957 that Hutchinson suggested a formal definition, clearly differentiating between what he called the "fundamental" niche and the "realized" niche.[5] The fundamental niche was the range of physico-chemical factors within

Figure 3.1 Vegetation frequently occurs in a non-random pattern of clusters. a. Labyrinthine
pattern in Western Namibia. b. Clusters of vegetation in wetlands of French Gui-
ana, originally emerging from farming activities of early Americans, but persistent
due to endogenous ecological forces.[6]

Source: (a) Nicolas Barbier: https://commons.wikimedia.org/wiki/File:Gapped_Bush_Niger_Nicolas_
Barbier.jpg (b) McKey et al., 2010, reprinted with permission from PNAS.

which the organism in question could conceivably exist, its various physical
limits. The realized niche, in contrast, was the position within the subspace of
the fundamental niche, where the organism did in fact exist. The realized niche
obviously includes effects from other species as well as potential effects on the
physico-chemical environment that the species in question might cause.[7]

A long-standing debate in ecology, still not completely resolved, is whether
organisms are generally limited by the physico-chemical limitations of their
fundamental niche (frequently referred to as density independent limitation) or
whether biological forces, either from the individuals of that species or other
interacting species, regulate the population into its realized niche (referred to
as density-dependent regulation).[8] This framework has been most frequently
associated with the population density of a species, but the formation of spatial
patterns has also been discussed within this same framework, and, we think, is
most usefully formulated in this way.

Imagine, for example that the distribution of individuals in a population presents itself as in Figure 3.2b. Clearly the distribution is non-random, indeed it is
"clustered" in the sense that individuals tend to be grouped together. But what
causes the clusters? Let us examine a couple of obvious examples. In the photo
in Figure 3.2a, we can see flocks of birds flying in the air over Rome forming
beautiful patterns called murmurations. Are those patterns caused by some
underlying template? That is not a difficult question. There is no underlying
structure that the birds sense and fly into to make these patterns, and it would be
very difficult to imagine some sort of template that stipulated where those clusters should be, independent of the birds themselves (recall the similar example in
Chapter 1). The pattern itself is "endogenous," the result of internal forces, not

of any outside influences, and it is only the behavior of the organisms involved that creates it. However, in Figure 3.2c we illustrate five patches of terrestrial vegetation. Obviously it is an aerial photo of the Hawaiian Islands, and it would be absurd to suggest that the island was formed by the vegetation – there is no biological process inherent to the vegetation that creates the particular patches of terrestrial vegetation. The pattern itself is "exogenous," caused by a pattern independent of the organisms that we observe having that pattern. These examples are clear. Spatial patches of birds emerge even in the apparent homogeneity of the air above Rome (Figure 3.2a). Patches of vegetation on the Pacific Ocean occur because the islands provide the only physical space where they can grow (Figure 3.2c). But the pattern in Figure 3.2b remains enigmatic without further information. Is there any way to tell if it is endogenous or exogenous simply by looking at it, or by measuring it in some way? The answer is probably no. However, there are some hints that one might use, some regularities that show up whenever a pattern is endogenous, and those regularities can be used as partial evidence of either endogenous or exogenous origin of the pattern. Later in the chapter we reveal the origin of the pattern in Figure 3.2b.

The sorts of non-random distribution of individuals in space illustrated in the depictions in Figure 3.2 are significant in two distinct, but interrelated, modes. First, when expanding into the fundamental niche (refer to Chapter 2, if necessary) and limited by that fundamental niche, there may be discrete habitats in which the individuals are able to live (as in the case of the vegetation on the islands), and those discrete habitats will dictate the spatial pattern that the

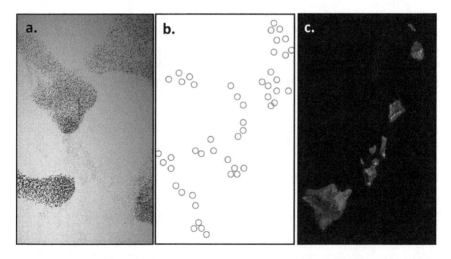

Figure 3.2 Clustered distribution of organisms. a. Birds flying above Rome (source: COBBS Lab, Institute for Complex Systems, CNR, Italy). b. An unspecified distribution of an organism, obviously clustered. c. Clusters of terrestrial vegetation on the Hawaiian Islands caused by the physical presence of the islands in the Pacific Ocean.

individual takes up. All population interactions are constrained to take place within that heterogeneous space. So, for example, if patches of anoxic voids in the rhizosphere contain various species of methanogenic bacteria, the composition of that bacterial community will be partly determined by the way in which the voids are colonized by particular species and their subsequent interactions. But the distribution of the voids themselves is independent of the actions of the bacteria. It is an exogenously generated pattern, and provides the background for the fundamental niche of the organisms that occupy it. Second, when specifically associated with the realized niche, which is to say organisms interacting with one another (see Chapter 2, if necessary), the biology of the species in question may cause patches, or some other pattern, to form. So, for example, the spatial pattern of a predaceous beetle may be a result of a combination of its eating rate and dispersal ability, but have little to do with any underlying physico-chemical limitations. It would thus be an endogenously generated pattern.

Thus, there are two aspects of non-random spatial pattern. First, there is a pattern effectively determined by the fundamental niche, in which spatial pattern is fixed completely independently of the activities of the organisms in question, but may nevertheless drive critical features of their population dynamics. Second, there is pattern determined through the realized niche, in which spatial pattern emerges from the population dynamics themselves, including any interactions with other elements of the biota. In one case, patches of biological activity are determined by underlying spatial pattern that is independent of those activities, while in the other case patches of biological activity emerge from the biological activity itself. Both of these issues can be important for many agroecosystem functions.

To some extent, this framing is a standpoint framing, which is to say the point of view we begin with in asking questions about spatial pattern matters. On the one hand we ask what might be the results of organisms entering an environment that is heterogeneous to start with. Here we are unconcerned with the origin of that heterogeneity, but rather seek to understand how it affects the population. On the other hand we ask what might be the origin of nonrandom spatial pattern, or, more specifically, do the organisms of concern have anything to do with creating their own spatial pattern. We begin with an analysis of the first (what are the dynamic results of background exogenous patterns) followed by an analysis of the second (how can patterns emerge from the endogenous dynamics of the organisms of concern), and finally follow with an analysis of the way in which the two forces may interact.

Dynamic consequences of background exogenous pattern

In consideration of the way in which pests arrive and occupy various agricultural fields, Levins noted that they were much like disease organisms – fields could be "infected" or not and the infected fields could "recover" if the farmer sprayed a pesticide or if a natural enemy arrived and drove the pest to extinction in that field.[9] Looking at the pest in this way, Levins suggested the revolutionary idea of formulating population dynamics as consisting of habitats occupied (field

a.   b.

Figure 3.3 Landscapes from Brazil's Atlantic forest region. a. Forest fragment next to a sun coffee farm. b. Forest fragment next to a degraded pasture.

infected with the pest) rather than the population density of the organisms (number of pest individuals in the area). Thus was born metapopulation theory.[10]

In most general terms, metapopulations are representations in which the proportion of available habitats occupied is the subject of study. They may be, for example, islands in the Pacific Ocean, mountaintops in western North America, forest fragments in Brazil's Atlantic forest region (Figure 3.3), or agricultural fields anywhere. The point is that the dynamics (i.e., change over time) emerge from a balance between the rate at which unoccupied patches become occupied and the rate at which occupied patches become unoccupied.

To take an artificial example, suppose we have 100 fields and 25 of them are occupied by a pest species this week. How many will be occupied next week? Elementary considerations suggest that if the probability that a successful colonist is constant over all occupied fields, the probability that a colonist will be somewhere in the overall region will be mp (migration rate, in this case m, times proportion of fields occupied, in this case $p = 0.25$). And the probability that one of those colonists will successfully occupy a previously unoccupied field is $1 - p$ (the proportion of fields not occupied, in this case $1 - p = 0.75$). So the overall probability that an empty field will be occupied by a pest is the probability a pest will become a colonizer or migrant, mp, times the probability that it will encounter an empty patch $(1 - p)$, or generally, $mp(1 - p)$.

Naturally, if the increase in fields occupied is $mp(1 - p)$, since that quantity is greater than 1 (p always is less than 1 – it is a proportion), the system will always increase to the point that $p = 1$ (as long as m is positive). However, we must also take into account the contrary tendency. Pests sometimes are eliminated from fields (spraying pesticides or runaway predation by natural enemies, for example). If we presume that the probability of a local extinction event is constant, call it e, the proportion of fields experiencing an extinction event during a week is ep. So the increase in occupations $[mp(1 - p)]$ will be balanced by the decrease due to local extinctions (ep) and the rate of change (positive or negative) will be:

Rate of change $= mp(1 - p) - ep,$

and at equilibrium, when the rate of change = 0, we can do a bit of algebra to see that,

$$p* = 1 - \frac{e}{m}$$

where the asterisk adjacent to the p indicates "at equilibrium."

Thus we intuitively see that the long-term occupation of fields in a large region will be dependent on the ratio of extinction to migration. When the two are equal ($e/m = 1$), $p*$ will be equal to zero and the population will go extinct over the whole region. It is also of interest to see that with these very simple dynamics we expect that only a fraction (precisely $1 - e/m$) of the fields in a region will eventually become infected, although which particular farms are infected will be ever changing. Thus, a question like "why is my farm infested but my neighbor's is not?" may not have any other answer than "you had bad luck." However, it also suggests that the way to reduce the average attack rate is (1) by increasing the average extinction rates, which might require individual farmers taking action to reduce the pest within their farms, or (2) by decreasing the average migration rate, which might require collective action of farmers and others on a regional scale to prevent pests from migrating from farm to farm. So, for example, if small-scale vegetable farms are distributed in a landscape where large scale sugarcane farms dominate, it could be that pesticide spraying in the sugarcane reduces the migration rate of the pest among the vegetable farms, effectively reducing m and therefore lowering the number of vegetable farms that will have the pest. However, it could also be that pesticide application in the sugarcane farms eliminates the natural enemies at the landscape level and thus the pest suppression by natural enemies within each individual farm will go down, which is to say, e for the pest will go down causing the number of farms infected with the pest at the landscape level to go up.

The metapopulation framework has been used extensively in questions of landscape dynamics. Indeed, one of our previous books, *Nature's Matrix*, has this issue as its central concern. Conservationists have long noted that most natural vegetation, especially in tropical areas, is concentrated in "island habitats" which are located in a "sea" of agriculture. As a generalization, the current world situation in terrestrial areas – especially in tropical areas – consists of landscapes that are effectively constructed of fragments of natural vegetation embedded in a matrix of agricultural activities. Using the metapopulation framework, a given species that lives in those fragments is not likely to survive unless it lives as a metapopulation. Since natural local extinction rates are generally large, fragment by fragment the species will go locally extinct (disappear from a particular fragment), eventually becoming extinct over the entire landscape, unless the migration rate among those fragments is high enough to prevent it. Using the basic metapopulation framework, since $p*$ (the equilibrium fraction of fragments occupied by the species in question) is equal to $1 - e/m$, as long as m is relatively large compared to e, the population will survive in the long run. But this means that we need to be especially concerned with the size of m, which is related to the "quality" of the matrix. And the quality of that matrix is dependent to a great degree on the nature of the agriculture that occurs in it.

The Levins metapopulation model is an example of a model that treats spatial distribution as an overall average, or mean. It is concerned with the local extinction rate on an average farm, and the average migration rate over the whole region. This sort of vision of the process is referred to as a "mean field model," and is frequently very useful as an intuitive device as much as an analytical tool. Yet sometimes, the very specific nature of the spatial pattern is not conveniently summarized as a mean field.

While the mean field approach is attractive for its simplicity and elegance in formulation, the precise way in which organisms disperse over a landscape sometimes makes the mean field approach less than ideal. The precise way organisms are distributed in a physical space matters! In a less-than-ought-to-be cited experimental study, Huffaker[11] created a spatial network of oranges in the laboratory and cultivated mites that eat fungus that grows on the surface of the orange, along with another species of mite that eats the fungivorous species. His results show clearly that the predator/prey pair, when isolated on a single orange, rapidly oscillate to extinction, as would be predicted by basic predator/prey theory in a non-spatial context (see our brief explanation in the caption to Figure 3.4). However, when Huffaker added a spatial component to his experiment

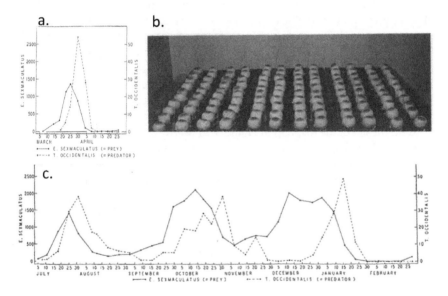

Figure 3.4 Huffaker's (1958) classic experiment. a. Predator/prey oscillations to extinction in a simple uniform environment. b. The experimental setup, a 10 × 12 array of oranges. Vaseline trails of various patterns were set up between the oranges, reducing the potential migrations of the mites (a mite that eats fungus on the surface of the oranges and a mite that eats the fungivorous mite). c. The evident persistence of both predator and prey over three predator/prey cycles that resulted from allowing sole level of migration of the mites from orange to orange.

Source: *Hilgardia*, University of California vol 27, 343–383, ©1958 The Regents of the University of California.

(by connecting the oranges in a way that allowed the organisms to migrate from one orange to the next), a surprising result emerged; both predator and prey persisted (with oscillations, of course) (Figure 3.4).

The way in which such a predator/prey system can operate in space is easily seen with a very simple thought experiment. Presume that over the whole region predator and prey will forever persist (neither will go completely extinct over the whole region – we will talk more about this assumption later). Then if we focus on one point in space it will either be occupied by prey or predator or will be empty (no prey or predator). We presume that prey and predator occupying the same point simultaneously will not last very long since the predator will simply eat the prey and then die from a lack of food. With such an arrangement we can see an inevitable cycle where an empty space becomes occupied by a prey (with a certain probability), then that space with the prey will be colonized by the predator (with a certain probability), and then that space will become an empty space once the predator eats the prey and dies from the lack of food (with a certain probability). In other words, an empty space will be replaced by a space with the prey, which will be replaced by a space with the prey and the predator together, which will be replaced by an empty space, and so on. This is effectively the pattern that produced the results of Huffaker's experiment. If the area where this is happening is small (like, for example, a single orange in Huffaker's experiment; Figure 3.4a), it is likely that the prey will not be able to migrate to an empty site (i.e., the probability that an empty site will be occupied by a prey is essentially zero), at which point the whole prey population will go extinct (and, of course, the predator, lacking food, will then die out also). However, if the area is much larger (Figure 3.4b and 3.4c), it is unlikely that all prey in the entire region will all go extinct at the same time, thus leading to the persistence of the three-state oscillatory sequence (empty, prey, predator, empty). We have diagrammed this process in Figure 3.5.

With the aid of the diagram in Figure 3.5 it is easy to see that if the feeding rate of the predator is so high that the prey becomes locally extinct (and, of course, subsequently the predator itself also, because of a lack of food), the only way the system can be made to persist is if there is sufficiently large migration rate of prey and predator, which obviously requires a distribution in some sort of geometric space. That is, when the site is empty, a migrant prey item fills it. But if that site

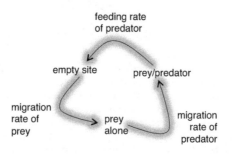

Figure 3.5 Qualitative representation of a predator/prey system in distributed space

is the only site in existence, there is no place the prey can come from. Thus there must be some other site from which the prey can migrate. The whole dynamic process in Figure 3.5 thus only makes sense when the system is distributed with some sort of geometrical extension in which each site is at least partially independent of each other site for some of the dynamic rules to play out.

But the other side of the coin is that the predator inevitably eats all the prey at a given site (Figure 3.5). Obviously this is not necessarily the case under all circumstances. However, especially in agroecological applications, it is commonly an important fact. Predators can, frequently, eat their prey so rapidly that the reproductive capacity of the prey cannot keep up with them, and the prey item (food for the predator) disappears, which leads to the predator itself also disappearing. Thus we see there are three dynamic processes operating in the simple framework presented in Figure 3.5. One of those processes operates at a local level (the rate at which the predator feeds on its prey), while the other two (the migration rates of the prey and the predator) operate in the framework of a larger spatial region. Indeed, the migration rates have no meaning unless the system is distributed in space. The existence of some sort of a spatial pattern thus determines the biological dynamics of the system.

Generation of endogenous pattern

Referring back to Figure 3.2, in addition to the dynamics of a population operating in an exogenously generated spatial pattern, recent years have seen the growth of interest in the spontaneous formation of discrete spatial patterns from dynamics that are internal to the population or system in question, endogenous pattern formation. An example for an understory woodland shrub is shown in Figure 3.6.[12]

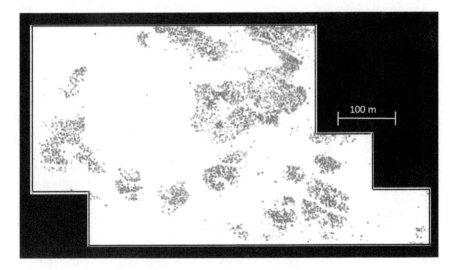

Figure 3.6 Distribution of witch hazel (*Hamamelis virginiana*) stems in the E. S. George Reserve (University of Michigan) in a 21-hectare plot

Note the obvious clustering of the trees. The background conditions – aspect, slope, soil, moisture – show no relationship to the clustering. What then causes it?

Another example comes from a study of nematodes in an agricultural field.[13] In this study, the field had been plowed yearly for many years, suggesting that the underlying physical and chemical habitat was probably well homogenized. Yet, in this presumably homogenized field nematodes in three different trophic groups showed distinct non-randomness, each group exhibiting a distinct pattern. Something other than the background conditions was creating this pattern.

These and many other studies that could be cited suggest that frequently there is a dramatic non-random pattern that emerges from the operation of the biological system itself, rather than from some preformed template. Elaborating on this theme, ecologists have provided a number of visions that could account for such pattern formation. One very simple formulation was elaborated by the British mathematician Alan Turing. It has been used as a conceptual tool to understand, qualitatively, how these sorts of spatial patterns may emerge spontaneously. Turing was thinking of developmental biology and how patterns such as spots on a jaguar's coat are formed. As we noted in our book *Coffee Agroecology*, there is metaphorical connection between the spots on a jaguar's coat and the spatial patterns of organisms on farms, or perhaps even farms in a landscape.

Turing's insight was motivated by the chemical reactions in developmental biology. He wondered what would happen when two chemicals react with one another in a concentration-dependent fashion. He postulated that chemical number one reacts with a solution of some other chemical to activate the creation of a product, which increases in intensity over time. But at some critical concentration, the reaction produces chemical number two, which suppresses the further formation of chemical number one. The two chemicals are referred to as the activator (first chemical) and the suppressor (second chemical). The key insight of Turing was to propose that the second chemical diffuses at a faster rate than the first one.

Suppose an automatic dispenser drips chemical 1 into a petri dish that contains a base solution. Chemical 1 combines with the base to produce a product that diffuses across the petri dish, causing the reaction not only where it was dropped, but also across the dish. But as the drops of chemical 1 increase, the local concentration of the product increases also as the activator chemical continues diffusing across the petri dish. There is a critical concentration of the product that stimulates production of chemical 2, which inhibits the interaction of chemical 1 with the base solution. But chemical 2 also diffuses across the petri dish faster and suppresses the formation of the product. The final result of this process is a fixed spatial pattern. We illustrate the process in Figure 3.7 and some actual Turing patterns that emerge in Figure 3.8.

Mathematical biologist Greg Murray has a fanciful, yet intuitively useful description of the Turing process that may make more sense than the chemically inspired explanation.[14] Suppose you have a field of grass. Ignite a fire in the middle of the field. As the fire burns (suppose it burns very slowly), it is

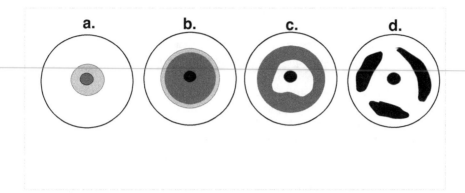

Figure 3.7 a. A petri dish with a solution (white) into which a drop of activator chemical is introduced in the center, reacting with the solution to create a product (dark gray circle in the center), and begins diffusing across the petri dish (light gray area). b. The product (central black circle) reaches a critical concentration such that a repressor chemical is produced (stippled area), which begins diffusing (the original activator chemical continues to diffuse – shading indicates recent versus old diffusion area). c. The repressor chemical cancels the activator creating a zone of original solution (white), but continues diffusing, although its concentration gradually diminishes as it diffuses. d. The final result is the formation of a spatial pattern.

Figure 3.8 Typical computer-generated Turing patterns

effectively the activator of another process, the regeneration of the grasses after the fire passes (we have to imagine an impossibly large field of grass to make this explanation work). The new grass shoots attract grasshoppers, who eat the new shoots but fly in every direction as they become satiated. But, and here is a particularly difficult biological assumption, the grasshoppers sweat a lot and wherever they go their sweat makes the grasses too wet for the fire to spread. So the grasshoppers act as the repressor agent through their excessive sweating. The activator (fire) and the repressor (grasshopper sweat) combine to create a pattern when the grasshoppers fly faster than the fire spreads.

It is worth noting that the Turing mechanism is a spatial example of a basic feedback process, one of the main ecological dynamical forces outlined in Chapter 2. An activator produces a product that "feeds back" on itself through the production of a repressor. The details of that feedback process are very important, especially the simple idea that the repressor must diffuse more rapidly than the activator.

One evident example from basic ecological processes is the so-called Janzen/Connell mechanism. Mainly applied to plants, but clearly of potential relevance for all other organisms, the idea is that a tree produces seeds, which germinate at some distance from the mother tree, creating a seedling shadow that is concentrated near the mother and declines in concentration as it recedes from the mother tree. But in a natural situation there are many natural enemies of those seeds and seedlings – seed predators, specialist herbivores, diseases. And generally it is thought that these natural enemies respond to the local density of their prey/host population.

Expanding on this fundamental idea, it is evident that any spatially distributed population that has natural enemies that respond to its local density, has at its foundation the underlying structure that could produce these basic Turing patterns. In the case of witch hazel (Figure 3.6), there is an explosive seed capsule and the seeds are expelled and fall to the ground relatively near to the parent tree. So when a group of trees is formed, the seeds that are ejected near the border of the group germinate and prosper, but those germinating within the dense shade of the central part of the group die. Thus there is a tendency for each cluster (group) to expand (this is equivalent to the activator in Turing's sense). Acting as the repressor is a beetle seed predator that builds up its population locally and disperses only rarely. On rare occasions, apparently, a seed from one cluster is distributed far away from the cluster, an isolated witch hazel tree emerges – several isolates can be seen in Figure 3.6. The beetle seed predator seemingly disperses only slowly, but when encountering a dense cluster of witch hazel plants, finds an abundant source of seeds to eat, and builds up a locally dense population, effectively stopping the further expansion of that cluster of plants.[15]

One agroecological example is the distribution of the predatory ant, *Azteca sericeasur* (formerly *A. instabilis*), in coffee farms in Mexico. This ant builds its nests in the shade trees of the coffee farms and it is thus easy to locate both potential nest sites (an individual shade tree) and the nests themselves (trees occupied by workers of this species are very obvious since they get angry and swarm all over you if you bang on the tree containing their nest). The actual distribution of nests in 2014 is displayed in Figure 3.9 where it is reasonably clear that the nests are mainly clustered in various sized groups. Since this ant species is well-known as a keystone species in a complicated network of pest control,[16] its particular spatial distribution is of some importance. Furthermore, given this importance, understanding how the spatial pattern emerges is likewise of practical interest. As we will discuss, the way that spatial pattern emerges is actually quite interesting, well

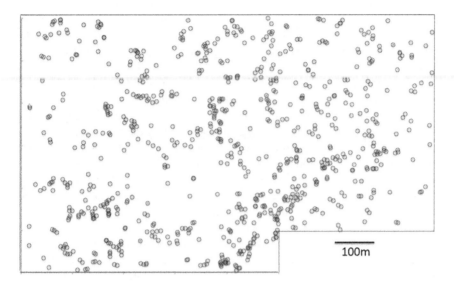

Figure 3.9 Spatial distribution of the ant, *Azteca sericeasur*, in a 45-hectare plot within a coffee farm in Chiapas, Mexico

beyond its practical applications, bringing to light some recent research in spatial ecology.

The *Azteca* ant is mutually associated with a scale insect (a pest of coffee), and that association is very tight, a system that we introduced at the beginning of this chapter. This mutualism is a typical ant-hemipteran mutualism where the hemipteran provides honeydew that the ants harvest and use as their main source of energy, while the ants protect the hemipteran against their predators and parasitoids. Anything that attacks the scale insect will have a negative effect on the ant since they rely on the honeydew for energy. In the *Azteca*-scale system in Mexican coffee farms there are two major enemies of the scale insects, a scale-eating ladybeetle and a fungal pathogen, both of which are "density dependent" on the scale (the more scales, the more beetles and fungus). When the scales are under the protection of the *Azteca* ants, their numbers build up significantly, potentially creating a problem for the coffee plant that harbors them.[17] Additionally a fly parasitoid directly attacks the ant, causing mainly behavioral disruption. The fly is also a density dependent enemy of the ant (there is a higher attack rate in large clusters of ant nests). When the fly appears, the ants stop foraging, causing a disruption of the protective function of the ants on the scale insects, and the presumed tendency of the ants to move their nest to a different location free of flies.[18] The ants have multiple queens in each nest, and as the colony grows it has a tendency to bud out and establish new nests in nearby trees. So, metaphorically, the ant, through its nest-budding behavior, acts as an activator, while any one or combination of the three evident natural enemies – the lady beetle and the fungus that attack the scales and the fly that attack the ants – may act as the repressor, resulting in a spatial pattern that we refer to as a Turing pattern.

It is worth noting that a pure Turing pattern is fixed in space while patterns such as the one just described are actually quite variable. Some authors will insist that the appellation Turing mechanism be applied only to those systems that correspond to the original equations of Alan Turing, which are reaction diffusion partial differential equations. Regardless, it is clear that spatial pattern formation results even in other modeling frameworks (e.g., cellular automata, as we employed originally for the *Azteca* ant data), and those patterns sometimes reflect remarkably the qualitative nature of the patterns originally predicted by Turing. In recognition of the basic insight of Turing, which is most important mainly as a qualitative insight in our view, we use the intuition gained from his mathematics to generalize. A system with some dynamic properties operative in a spatial context wherein there is a component that activates and another component that represses and the dispersal of the activator is slower than the dispersal of the repressor, is expected to form a spatial pattern, and we refer to those patterns as Turing patterns.

Returning to the beginning of this chapter, distinguishing between exogenous and endogenous driving forces is sometimes not so simple as finding spatial correlations. Consider, for example, the examples of vegetation patterns in arid and semi-arid areas (as in Figure 3.1a). According to some authors,[19] these patterns are caused by the dynamics of water utilization (Figure 3.10). We can see how the Turing mechanism can aid the intuition here – if the root system expands faster and further in space than the shoot system, a zone of water deficiency will be created at the edges of patches, which will make those patches unavailable for new seedlings to become established.

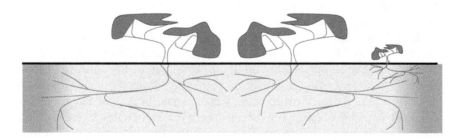

Figure 3.10 A proposed mechanism for generating vegetation patterns in arid and semi-arid environments. The root systems grow more extensively than the aerial vegetation, thus producing zones of effective water (and contained nutrients) depletion where new seedlings cannot become established (Rietkerk et al., 2004).

Criticality and power functions

When organisms "self-organize" according to a particular rule that is independent of the external environment (i.e., when the pattern formed is endogenous, or based on the realized versus fundamental niche), it is frequently the case that they form clustered patterns, much like the ants in Figure 3.11, either the real ants (Figure 3.11a and 3.11b) or a computer projection of them (Figure 3.11c and 3.11d). Yet many forms of clustering may exist, which is to say, spatial distributions may be non-random in a variety of ways. For example, in Figure 3.12 we illustrate first a random pattern (Figure 3.12a), then a highly clustered pattern where the clusters have a central tendency (i.e., all clusters tend toward a common size, a mean of about eight nests per cluster – Figure 3.12b), then the clusters of *Azteca* ants as they existed in the 45 hectare plot within the coffee farm in 2010 (Figure 3.12c). In other words, it may not suffice to ask simply whether the spatial distribution of an organism is random or not, but may sometimes be more important to characterize a non-random distribution more carefully. What sort of non-random distribution do they follow?

Field observations (two different years)

Predictions from computer model

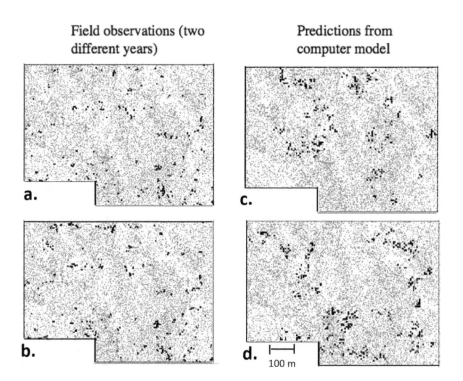

a.

b.

c.

d.

100 m

Figure 3.11 Field observations of the distribution of *Azteca sericeasur* nests in 2004 (a) and 2005 (b) and results of computer simulations with a Turing-based spatial model for the same data (c and d).[20] Small grey points are the approximately 11,000 trees in a 45-hectare plot and the dark symbols are the trees that contain an *Azteca* nest. Note the plot shown in Figure 3.9 is the same plot a decade later.

a.   b.   c.

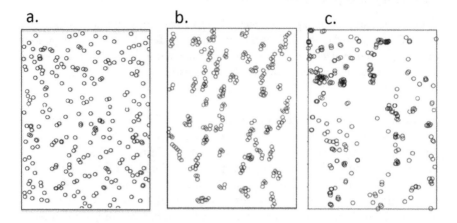

Figure 3.12 Example distributions of ant nests. a. A completely random distribution. b. A
        non-random clustered distribution, with a central tendency (average cluster size
        is about eight). c. The actual distribution of *Azteca* ant nests in a 45-hectare plot
        within a coffee farm in 2010.

In a modestly titled book called *How Nature Works*, physicist Per Bak motivated
many disciplines to begin thinking of particular patterns in nature and how they
were formed.[21] His classic example is a sand pile. Suppose, he said, you have a
funnel, and you slowly drop sand into it such that the sand accumulates on a base
below (Figure 3.13). The accumulated sand on the base eventually takes on a
form that looks like the basic shape of a volcano, at which point its overall shape
does not really change any further, even as you continue pouring sand into the
funnel. It has, in a sense, "self-organized" itself into this easily recognized shape.
What is amazing about this process is that once the pile reaches this basic shape,
not only does it not change shape in the future, but the sand grains that fall off
of the side of the pile as more are added to the top show a very regular pattern.
Each new sand grain that falls on the top of the pile causes a small avalanche of
sand grains somewhere on the side of the pile. If you carefully count the number
of sand grains in each avalanche, you come up with a remarkable pattern. As
we see in Figure 3.13, the size of the avalanches (the number of sand grains in
an avalanche) follows a power function (if $y(x)$ = proportion of avalanches of
size x, then $y(x) = ax^b$ such that a plot of the log of $y(x)$ against the log of x will
be a straight line, as shown in Figure 3.13b). The equation of this straight line,
$[\log(y) = \log a + b[\log(x)]]$, makes clear the significance of the two parameters (a
and b) of the power function. As can easily be seen in Figure 3.13b, parameter
(b) is the ratio of the log of the frequency of the singletons (clusters containing
a single individual) divided by the log of the size (number of individuals) of the
largest cluster, that is, the slope of the line. In other words, the log of the total
number of tiny avalanches (those that have just a single sand grain) divided by

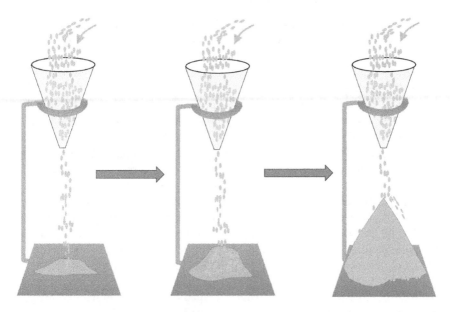

Figure 3.13 Example of a self-organized critical system, the sand pile. Beginning at the left, sand is dropped into a funnel and slowly streams out the bottom, creating a small sand pile. The middle graph shows the sand pile growing until, at the right, the sand pile reaches its ultimate form (shaped like a volcano), at which time each sand grain that falls on the top creates an avalanche of sand grains that tumbles off the sides.

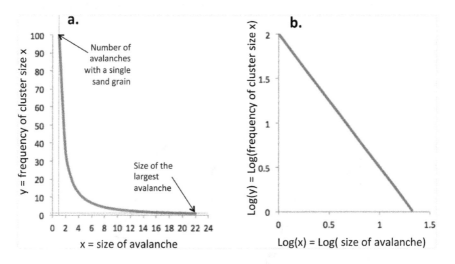

Figure 3.14 The relationship between frequency of clusters as a function of cluster size, often described with the equation $y = ax^b$, where y is the frequency and x is the cluster size. a. Plot of the basic equation, emphasizing the intercepts when $x = 1$ (a cluster size of one individual) and $y = 1$ (when there is a single cluster of the maximum size). b. A double log plot of the same variables, reflecting the linear form of the equation, $\log(y) = \log a + b[\log(x)]$, wherein we see the meaning of the two parameters of the power function.

the log of the number of sand grains in the biggest avalanche, is the parameter of the power function. The parameter (*a*) is simply the exponentiation of the frequency of the singletons. While there is some debate about the meaning of power functions in the literature, it is frequently taken as a hint that some sort of self-organizing process is involved.

Much as the distribution of avalanches in a sand pile form a power function, the sizes of clusters of *Azteca* ant nests on a coffee farm also do, both from a theoretical model formulation (explained further in Chapter 8), and the empirical data. In Figure 3.15 we show that fit for the distributions displayed in Figure 3.11.

Perhaps an approximate metaphor will prove useful at this point. Suppose we are at the frontier of an agricultural expansion zone. New farmers clear land at some particular rate. Suppose that farmers' attitudes range from those who want to be isolated and not bothered by pesty neighbors to those who feel the need to be part of a more connected community. From this situation you can imagine the formation of a spatial pattern of farms. There will be some clusters of farms, with some farms isolated and alone, depending on the nature of the attitudes of the farmers. There is something special about the situation where the total number of isolationist farmers is equal to the total number of socially connected farmers (the isolationists will tend to keep their farms unconnected to neighbors, while the socially connected ones will tend to group together into a single large cluster of farms). While we cannot say for sure whether or not the farm clusters will be distributed like a power function, we can say that if they indeed do follow a power function, the parameter (*b*) of that function will be 1.0 (since the parameter is the ratio of the log of the number of isolated farms to the

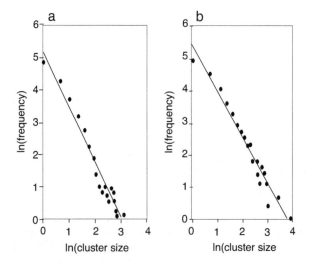

Figure 3.15 Power function fits for the distribution of ant nests in Figure 3.12. a. From the natural distribution. b. From the computer projections.[18]

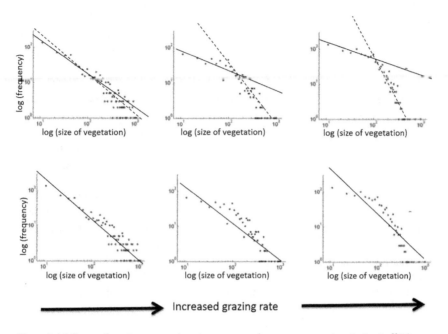

Figure 3.16 Power function approximation to natural pasture vegetation in Spain.[22] The top row shows two approximate fits to a power function at three different grazing densities. The solid line uses only low values of the vegetation patches. The dashed line uses only large sizes of vegetation patches. The bottom row is a power function fit to all the data, illustrating the systematic nature of the deviation from the power function occurs as grazing rate increases, wherein there are increasingly too few of the smaller patches and the largest patch is too small.

number of farms all clustered). Interestingly, the power function form enables an intuition about spatial pattern, an example of which is shown in Figure 3.16.

In a spatial pattern in which individual elements are the members of spatial clusters, and where the total number of elements that are not part of any cluster at all (the singletons) is precisely equal to the total number of elements in the largest cluster (on a log/log plot, the intercept on the y-axis is equal to the intercept on the x-axis) the parameter (b) of the power law equation is equal to negative 1.0. Deviations from −1.0 can then be qualitatively understood as either "too many or too few singletons," or "the largest cluster is either too large or too small." So, given that clusters are formed in space, and that they do indeed seem to approximate a power function (look linear on a plot of the log of frequency versus the log of cluster size), it is of some interest to ask in what direction does the critical power function parameter of 1.0 deviate? If it is greater than one, that suggests that either the number of singletons (the smallest cluster size) is too large or the size of the largest cluster is too small. If the parameter is less than one, that suggests that either the number of singletons is too small or the size of the largest cluster is too large. Where one goes from that point to further interpretations and understanding of the system depends on the precise system under consideration. Yet the interesting fact that there are expected patterns for which the parameter is precisely equal to 1.0 is only one key parameter to examine. And it is perhaps not as interesting as the simple fact that the distribution follows a power function in the first place.

We can utilize the fact of a significant deviation of real data from a formal power function to enable some understanding of what underlying pattern forming processes might mean. We illustrate these deviations in Figure 3.16, for Spanish vegetation patterns in a natural pasture.[22] As intensity of grazing increases, the spatial pattern of the vegetation patches (composed of low-lying brush) deviates from the power function systematically, where the number of singletons is fewer than expected while the size of the largest patch is smaller than expected. Might we expect that such a deviation from the power function could be taken as a measure of excessive overgrazing? Might that deviation indicate nearness to some sort of critical condition whereby the vegetation will itself disappear? We return to this issue in Chapter 9. For now we note that if the parameters of a Turing process change, the resultant spatial pattern itself changes, and such change could be reflected in the fit to a power function (Figure 3.17). A rich and yet unresolved theoretical literature has evolved concerning this issue.[23]

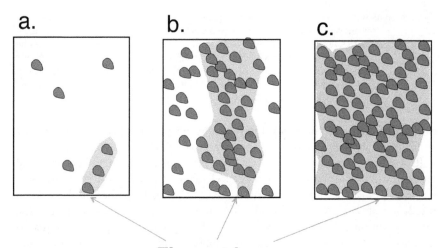

Figure 3.17 Diagram of the forest fire model, where shaded area indicates the spread of the forest fire. a. The density of trees is very low, so a fire initiated with a tree at the edge of the plot only percolates through a small number of trees. b. At a higher critical density, enough trees are in the population so that a fire started at one edge percolates all the way to the other edge of the forest. This situation is referred to as the percolation threshold and represents the point at which the trees that can ignite one another "span" the length of the forest. c. Beyond the percolation threshold, any fire percolates over almost all the forest area.

Percolation points and power functions

For certain spatial processes, there is a relationship between a close fit to a power function and the nature of the largest cluster in the spatial pattern. As an example, consider the model system that we reported earlier in this chapter, the formation of clusters of *Azteca* ant nests in a coffee farm in Mexico. The computer model referred to was a cellular automata model in which the spatial framework was likened to a chessboard with distinct cells (spaces) that could be occupied or not by an ant nest (in this particular case there were about 11,000 cells (to simulate the 11,000 trees in the 45 hectare plot that we sampled). Then a series of rules were placed on the system, in particular four rules that operated at each time step (from one year to the next):

1   An empty space would be occupied by the ant with a certain probability (the regional dispersal rate, meaning a new ant queen would arrive at a spot from outside of the array of cells).
2   An occupied space would become empty with a certain probability (the local density independent mortality rate).
3   An empty space would become occupied by the ant in proportion to the number of neighboring spaces that were occupied (the density-dependent local migration rate).
4   An occupied space would become empty in proportion to the number of neighboring spaces that were occupied (the density-dependent local mortality rate).

The idea behind number 4 is that there is a natural control agent that keys in to the local density of nests (in this case originally thought to be the parasitic fly that attacks the ants according to the local density of ant nests). The model system corresponds at least qualitatively to Turing's basic insight in that the nests expand locally (rule 3 provides the activator force), but are attacked by a natural enemy in proportion to local density (rule 4 provides the repressor force). The computer projections illustrated in Figure 3.11c and 3.11d resulted from this simple model.

The idea of percolation has been an important idea in many different branches of science for many years.[24] Perhaps the most obvious example is the idea of a forest fire. If we imagine a forest landscape with some density of trees randomly distributed, it is reasonably clear that if the density is very low, a fire on one or two trees will not spread very rapidly. However, if the tree density is very high, a fire started at one edge of the forest will "percolate" all the way to the other edge of the forest. The idea is illustrated in Figure 3.17.

If there are only a few trees in the forest (Figure 3.17a), a fire started at one edge of the forest will penetrate only a small distance into the forest. At the other extreme (Figure 3.17c), with a very high density of trees, a fire started anywhere will likely percolate throughout the entire forest area. There is a critical point where the density of trees is such that the fire will span the forest from one end to the other (Figure 3.17b). This density of trees, where a "spanning cluster" is generated, is referred to as the percolation threshold, or the critical percolation point. It turns out that the frequently observed power law of cluster distributions is intimately related to this idea of a spanning cluster, at least in some circumstances.

As an example, our simple cellular automata model for the *Azteca* ant can be likened to the forest fire in the sense that each square in the chessboard is either occupied by an ant nest or not. As the population density of ants increases, much as the population density of trees increases in the forest fire model, there is a critical point at which one of the large clusters spans the whole arena. It is precisely at this point, the emergence of a spanning cluster, that the distribution of clusters precisely follows a power function. Thus, for example, as we decrease the density dependent mortality rate (see the earlier description of the system), we expect the population density of ant nests to increase since there will be fewer "deaths" of nests, all other things held constant, certainly not a surprising element of basic population ecology. However, the context here is in a spatial extension of the population, and thus in addition to the general population density, we are interested in the spatial pattern that emerges. In Figure 3.18 we illustrate the interrelationship among the four factors discussed thus far: population density, spatial pattern, emergence of a spanning cluster, and fit to a power function.[25]

In this particular pattern-forming framework we see that, unsurprisingly, as the mortality rate increases the population declines (it declines at small values also, even though it is difficult to see on the log scale, which we use to emphasize the point where the population collapses). When the density-dependent mortality rate is extremely high (relative to the other processes in the model), the population cannot be sustained. At the point where the population is sustainable, its clustered pattern in space appears rather sparse, and the fit to the power function is biased, with "too few" singletons (single isolated cells occupied) and a maximum cluster size that is "too small" (see graphical insets to the right of the dashed line in Figure 3.18). Recalling the previous example of rangeland in Spain (Figure 3.16), it is the same as the pattern observed at higher levels of grazing, and here we know (since it is a model) that the general deviation from the power function does indeed provide us with a signal of impending population collapse. As the mortality rate decreases, the overall population density increases and the spatial pattern appears less sparse but, most importantly, the basic deviation from the power law (fitting the function to all of the data) remains the same – too few singletons and a largest cluster that is too small. Yet the deviations at both ends of that spectrum are less pronounced. At precisely the point where a spanning cluster emerges (vertical black dashed line at about 0.3 mortality rate in Figure 3.18) the cluster distribution follows a power function almost perfectly. As the mortality rate is decreased further, the spanning cluster tends to merge with other large clusters to form an extremely large cluster (see the single points on the x-axis of the two graphs to the left of the vertical dashed line in Figure 3.18, noted by the thin dashed arrows). The deviation from the power law thus is quite different

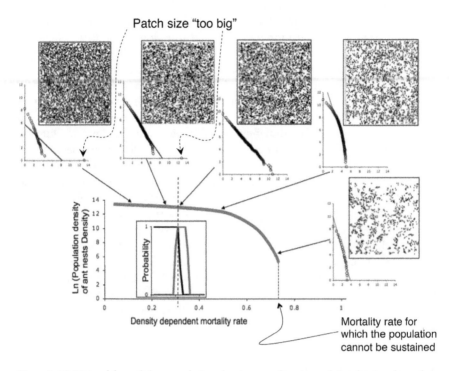

Figure 3.18 Natural log of the population density as a function of the density dependent
local mortality rate (red curve). Five examples from five different values of the
mortality rate along with spatial pattern for the "standard model" (see Vandermeer and Jackson, 2015) on a 100 × 100 lattice are shown. Vertical black dashed
line indicates the value of the mortality rate for which a spanning cluster first
appears. Note the single very large cluster in each of the two upper left graphs
(the outlying point to the right on the x-axis of the log/log plot, indicated by the
thin dotted arrows), to the left of the percolation point. The insert graph is the
probability of having a spanning cluster (black curve) and probability of a power
function fit being larger than 0.95 (red line), plotted on the same base mortality
rate axis. If we take the critical value of the probability of fit as 0.95, a deviation
from the power law cannot be assured for a broad range of local mortality, which
is the signal of "robust" criticality.[26]

than the earlier forms (to the right of the spanning cluster in Figure 3.18). Here
the largest cluster is too large and there are too many singletons.

A qualitative interpretation of the preceding model analysis suggests a useful
way of interpreting spatial pattern dynamics. Begin with a focus at the point
where the first spanning cluster emerges (in the case of Figure 3.18 that point is
where the mortality is approximately 0.3). That is a point at which the spatial
system indeed follows a power law, and can be thought of as a kind of canonical form. With lower mortality (and thus, higher population density), the larger

clusters begin to merge with one another and the deviation from the power function takes on a particular form (largest clusters too large). Moving in the other direction, increasing mortality rate (and thus, lower population density), the power function takes on a different form, with lower than expected singletons and smaller values for the largest clusters. In all cases it is true that there is a non-random distribution of clusters, as implied by the basic Turing framework. However, that non-randomness takes on further structural details, some of which can be envisioned by an examination of the deviations from the theoretical canonical point (power law distribution and single spanning cluster). Thus, for example if the organism of concern is a particularly effective nitrogen fixing bacterium and the matrix is a traditional maize field, if there are bacteriophages that consume that bacterium at different rates (say one is a major predator and decimates the population locally, but does not disperse very widely, while the other is relatively benign, but disperses throughout the field very efficiently). If the bacteria exists with one main cluster that is very large, that may imply that the "efficient" bacteriophage will have a major negative effect on the bacteria population. Contrarily, if the pattern formation mechanism produces a deviation from the power function such that there are "too many" small patches (a power function slope too steep, i.e., less than -1.0), the noxious bacteriophage may not be able to survive due to its inability to efficiently locate the isolated patches of bacteria.

The skeleton of the fundamental niche, the flesh of the realized niche

We started this chapter by suggesting a parallelism of the fundamental niche with exogenous spatial pattern formation and the realized niche with endogenous pattern formation (in Figure 3.19, as promised, we reveal the underlying structure that gave rise to the pattern in Figure 3.2b). However, it may be the case that many patterns emerge neither from the underlying fundamental niche nor from the realized niche, but rather from an integrated combination of the two. A remarkable example of this phenomenon comes from a situation in which agricultural activities sets the stage (the fundamental niche background for a suite of organisms), and soil "engineering" organisms reinforce the pattern, maintaining structure over long time periods, the basic dynamics that led to the vegetation patterns shown in Figure 3.1b. Early Americans initially created a system of raised fields, small regular patches of soil that stand above the surrounding wetlands, and cultivated them for some time. It is not uncommon for raised field agriculture to emerge in wetland situations,[27] although the size of these particular platforms is unusually small. After abandonment it would normally be expected that the elevated platforms would erode away after some years of seasonal rainfall and drainage. However, in this case the platforms originally set up by traditional agriculturists receive enormous inputs from

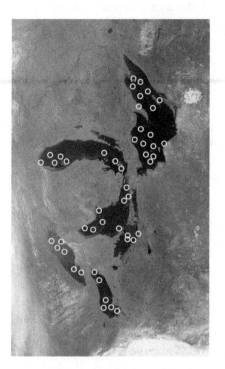

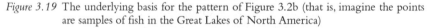

Figure 3.19 The underlying basis for the pattern of Figure 3.2b (that is, imagine the points
are samples of fish in the Great Lakes of North America)

Source: Based on image from SeaWiFS Project, NASA/Goddard Space Flight Center, and ORBIMAG.

ecosystem engineers, such that their basic structure seems to be maintained,
probably in perpetuity.

Two of the main engineers that act to perpetuate these structures are the
fungus cultivating ant *Acromyrmex octospinosus* and earthworms. The ant nests
almost exclusively on the ancient raised beds. The organic matter brought to
the raised bed by the ants seems to attract large numbers of earthworms, such
that during the dry season the surfaces of the raised beds are completely covered in earthworm casts, structures found exclusively in the raised beds. This
accumulation of organic material on the raised beds provides a background
for the growth of woody vegetation that would otherwise not grow in the
swamp, putting roots into the platform that provide for soil aeration and drainage and resist the expected erosion of the platform. Thus, this system is one
in which there is a sort of skeleton of structure formed by early agricultural
activities, and this structure is maintained subsequently by "self-organizing"
natural processes.

Spatial pattern and the agricultural connection

We opened this chapter with our observations on a Brazilian cacao farm, noting that the subtle consequences of pattern formation of a species of hemipteran-tending ant could be important for pest control, even if not clear from casual observation. This sort of spatial complexity can be confusing, to be sure. In trying to make general sense of the whole complicated issue of spatial pattern in ecology, as we think it applies to agroecosystems, we position our presentation within the standard framing of fundamental versus realized niche. The fundamental niche, in this framing, provides a background spatial pattern that dictates the spatial pattern of the organism that fits into it. The resulting dynamics of the organisms is sometimes dramatically influenced by that pattern – it could be the arrangements of farm fields in a landscape or the pattern of olive trees in an olive grove that makes populations of beneficial fungi operate as if they were metapopulations or olive scale insects act as they are source sink populations. The very fact of spatial heterogeneity that is structured within the fundamental niche can, as happened in the case of Huffaker's famous experiments, determine such basic features of ecosystem dynamics as to whether the system will be dominated by beneficial organisms or noxious pests.

Yet, there is a less well-appreciated aspect of spatial dynamics, the general subject of "self-organization," in which the organic interaction of the components of the system is determinate of the spatial pattern, and thus effectively a component of the realize niche. Our approach takes Alan Turing's basic insight as a qualitative framing, incorporating a variety of what some other authors might dissect out as alternative mechanisms. Although we are sure there are some who will differ with our decision to partake of this framing, it seems to us more important to ask what are the consequences of this spatial patterning. A common analytical framing is to use the fit to, or deviation from, a power function as a measure of spatial pattern. The deviation can be thought of as either "too many small patches" or "the largest patch has too many elements" (of course the reverse of each of these is also interesting).

We have argued elsewhere[28] that it is frequently the case that the autonomous generation of spatial pattern of one organism creates the habitat pattern that some other organism experiences. In this sense the self-organized pattern of species A may form the "fundamental niche" for species B. If this be the case, we can say something about the ultimate generation of at least this component of the fundamental niche of species B. If that niche contains one very large patch (which means that the dynamics of species A has produced a spatial distribution with a very small negative value of the power function parameter), it is likely that species B will exist as a source sink population (the biggest patch of species A being the habitat for the source population of species B). This context may at first seem strange since the fundamental niche is normally thought of as relating to the physico-chemical elements of the habitat. But our approach sees the fundamental niche as simply independent of the dynamics of the organism of concern. So, for example, big cat predators are more efficient hunters in low grass habitats than in high grass habitats. From the cat's point of view, it matters little whether the height of the grass emerges from underlying soil chemistry or competitive interactions among grass species.

The biocontrol beetle we introduced in the introduction to this chapter is an excellent example. The scale insects (the food of the beetle and a pest of coffee) appear to survive the dry season within concentrated patches of their mutualist, the *Azteca* ant, from which they disperse during the wet season, thus effectively living in a source sink population, determined by the *Azteca* ants, and the repressor elements in the system. The large patches of Azteca ant nests form the habitat that allows the source population of scale insects, and the isolated smaller patches of ant nests form the habitats for the sink populations. The beetles take advantage of this basic structure by finding the concentrations of scale insects to locate their larvae. Without the basic pattern formed by the ants and their scale insects, the natural control of the beetle would likely be lost.

Finally, we do not wish to leave the impression that it is only the on-farm micro-spatial scaling that is relevant. The distribution of farms themselves or, indeed, the distribution of non-farm habitats, may be key determinants of regional nutrient cycling, pest and disease dynamics, and even economic prosperity. Further considerations of a sociological or political nature may provide insights at this higher level or organization.

Notes

1 This narrative emerges from a variety of studies of ourselves and other laboratories, especially that of Stacy Philpott, and is summarized in our book *Coffee Agroecology*.
2 Excellent summaries of this early ecological prospecting can be found in Tilley, 2011; Anker, 2009.
3 Ettema and Wardle, 2002; Tilman and Kareiva, 1997; Huang and Diekmann, 2003; Levin et al., 2012.
4 Refer to the discussion of this issue in Chapter 2, also see Elton, 2001; Grinnell, 1924.
5 Hutchinson, 1957.
6 McKey et al., 2010.
7 See the detailed description in Chapter 2, especially Figures 2.9 and 2.10.
8 Density dependent versus density independent regulation was a hot topic in the 1970s. For an alternative perspective, the density vague idea of Strong, 1986, provides an interesting alternative framing. The emergence of the theory of chaos put yet an alternative interpretation on the issue, as discussed in Chapter 5.
9 Levins, 1969b.
10 Metapopulation theory has since seen a burgeoning literature, summarized in Hanski, 1999.
11 Huffaker, 1958.
12 Allen, 2012.
13 Robertson et al., 1995.
14 Murray, 1990.
15 De Steven, 1982; Allen, 2012.
16 Vandermeer et al., 2010; Perfecto and Vandermeer, 2015a.
17 Jackson et al., 2009; Liere et al., 2012.
18 Vandermeer et al., 2008.

19 HilleRisLambers et al., 2001; Rietkerk et al., 2004.
20 Vandermeer et al., 2008.
21 Bak, 1996.
22 Kéfi et al., 2007.
23 Pascual and Guichard, 2005; Kéfi et al., 2011; Vandermeer and Jackson, 2015.
24 Stauffer and Aharony, 1994; Sahimi, 1994.
25 This model is extensively discussed in Vandermeer and Jackson, 2015.
26 The idea of robust criticality was introduced by Pascual and Guichard in 2005, and Vandermeer et al. claimed to have found an example in 2008. Essentially it is a phenomenon where a signal of criticality (e.g., a power function fit) extends beyond parameter values that generate a spanning cluster. In the case of *Azteca*, the distribution of nests seemed to be very far from representing a spanning cluster, yet their frequency distribution seemed to fit the power function quite well.
27 Crews and Gliessman, 1991; Renard et al., 2012.
28 Jackson et al., 2014.

4   Chaos

Introduction

There has been prodigious press surrounding the idea of chaos. The popular imagination is naturally stimulated significantly by a theory that seems to say it is impossible to understand anything at all. Indeed, in the proper framework, that is precisely what chaos suggests, and it is profoundly upsetting. Yet perhaps it is less unsettling when we understand that its essence is familiar to all of us in our shared common experience. Consider, for example, the process of making tortillas. After grinding the corn into corn flour, it is possible to identify small grains within the flour. Suppose, for some crazy reason, you want to know where a particular grain of corn flour will be after you mix the flour in water to make the dough that you will pound into a tortilla. If we were dealing with, for example, two flour particles on the surface of the dough, it is reasonable to expect that any two particles close to one another on the surface of the dough, will generally remain close to one another as the dough ball is pounded into a tortilla (Figure 4.1a). It almost seems too obvious to point out.

But if we do the same exercise, locate two particles of flour, before we add them to the water to make the dough, as we mix the flour into the water to make the dough, it is highly unlikely that those two particles will remain close to one another, even if they were extremely close (even touching) at the time when they were first poured into the jar of water (Figure 4.1b).

Flattening the dough ball to make a tortilla was like a non-chaotic system: Points close to one another on the dough ball remain close to one another on the tortilla. Mixing the flour into the water was like a chaotic system: points in the flour/water mixture will be completely independent of where they were when we started the process, that is, when we poured the flour into the jar of water. This is the formal essence of chaos, the idea that two starting points, no matter how similar they may be, will eventually deviate from one another a great deal in such a way that information about the one leads to no information whatsoever about the other. It is a condition known as "sensitive dependence on initial conditions."

With this intuitive insight into the nature of chaos, consider a more formal example. Suppose the rate of population increase of the bacteria in the soil of an

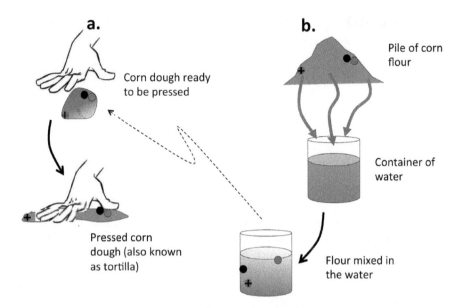

Figure 4.1 Illustration of the principle of "sensitive dependence on initial conditions."
a. The dough ball contains three particles, one red (light shading), one black (dark shading) and one green (with a small cross) on its surface, and when the dough is directly flattened, the particles remain more or less positioned with respect to each other the way they started. So, if we see that black and red in the tortilla are close to one another and both far away from green, that suggests that black and red were indeed close to one another in the original dough ball. b. Again there are three particles located in a pile of corn flour, black and red close together, and green far from them. After mixing in water, the information about the relative positions of those three particles contained in either the container of water or the container of the newly formed dough provide no information at all about the relative position of the particles at the start.

urban garden is dependent on the availability of nitrogen. But there is a complicated relationship between nitrogen and bacteria, such that when the bacteria get too abundant, many of the bacterial cells cannot get enough nitrogen to survive and there is a massive die off of bacteria cells. So, we have a pattern of increasing bacterial sequestration of nitrogen and increasing bacterial population, eventually leading to a collapse of the bacterial population due to the overconsumption of nitrogen (there is a formal mathematical model of this presented in the caption to Figure 4.2 – for now, you need not understand the model itself, only the results as we present them here). Suppose we have two fields, close to one another, but separated by a hedgerow or some other vegetation that keeps migration of bacteria from one field to the other to a minimum.

In the spring, growth is slow and the population increases slowly, eventually reaching a threshold where it stays for a long time (see Figure 4.2a and 4.2b).

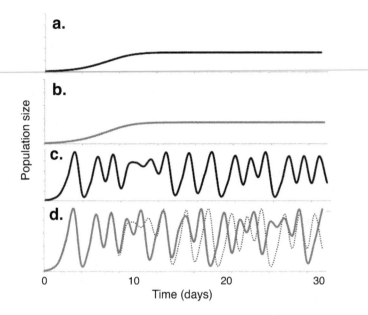

Figure 4.2 Illustration of population dynamics of two bacterial populations in spring (a and b) and summer (c and d). Note that in (d) the trajectory of (c) is plotted with a dotted line, illustrating how the two populations deviate from one another even though they began at almost identical initial conditions, an example of "sensitive dependence on initial conditions." On a technical note, the model is the logistic equation $N(t + 1) = rN(t)[1 - N(t)]$, where $r = 2.5$ for the spring population and $r = 3.9$ for the summer population.

Then, in the summer, because of warmer weather the bacteria reproduce much more rapidly, depleting their nitrogen supply such that their population collapses to almost zero, only to rebound when nitrogen stores are replenished. If we simulate the population growth using a simple population growth model of these bacteria, the pattern might be something like in Figure 4.2c. If we run the same model using exactly the same parameter values, which will be the equivalent of comparing the two fields, something interesting, and somewhat disturbing, happens. If we begin with almost exactly the same number of bacterial cells and the same amount of nitrogen in both fields, for a time the two populations behave in an almost identical fashion (they are almost identical from 0 to 10 days), but then begin to deviate, and eventually have completely distinct patterns of behavior (in Figure 4.2d we plot both populations together, repeating the population of Figure 4.2c with a dotted black line). If it would be desirable to know the precise population density sometime in the future, there is no way we could estimate the current population precisely enough to make that calculation, even if we have 99.99999% perfect knowledge of the population processes of birth and death.

That is, almost perfect knowledge about the system will not enable us to predict its future time course!

This bizarre situation was first recognized by the great French mathematician Henri Poincaré when he contemplated what would happen if we applied Newton's law of gravity to the sun, moon and earth simultaneously. As most people understand, Newton's laws predict very precisely the way the earth revolves around the sun and the way the moon revolves around the earth. But what surprises most people is that if we apply those same laws to all three bodies together, those bodies, according to Newton's laws, behave in completely unpredictable ways. Poincaré noted this in 1887, and for this reason he is sometimes thought of as the father of chaos theory. But his insight about the so-called three body problem was soon generalized by him to something that might be thought of as a general law of nature. In 1903 he noted,

> If we knew exactly the laws of nature and the situation of the universe at the initial moment, we could predict exactly the situation of that same universe at a succeeding moment. But even if it were the case that the natural laws had no longer any secret for us, we could still only know the initial situation approximately. If that enabled us to predict the succeeding situation with the same approximation, that is all we require, and we should say that the phenomenon had been predicted, that it is governed by laws. But it is not always so; it may happen that *small differences in the initial conditions produce very great ones in the final phenomena.* A small error in the former will produce an enormous error in the latter. Prediction becomes impossible.

Perhaps the most famous of recent elaborations of chaos theory comes from trying to understand the immensely complicated weather system, something farmers have been trying to do for ages, with some, but limited, success. We tend to think that the weather is so hard to predict mainly because there are so many factors to be considered – wind speed, temperature, moisture and so forth. Yet it turns out that such a multitude of factors may not really be the culprit. In 1979 a clever meteorologist, Ed Lorenz, decided to try and gain some insight about how weather systems behave by studying in great detail a small component of them, the convection roll. It had long been known that as air heats up near the surface of the earth it rises, but then begins cooling and at some height above the earth begins to "roll" down again. This process is known simply as a convection roll. Lorenz thought he might gain some insight through the study of a simple convection roll. A classic set of equations had already been devised using three basic parameters (the rate of rising and falling of the air, the difference in temperature between rising and falling air, and a measure of the deviation from straight up and down of the roll). In one of those episodes in the history of science where surprising results change our interpretation of the world, Lorenz discovered that this extremely simple model of the weather system (or, rather, a very small component of the weather system) was chaotic. That is, the notorious unpredictability of weather systems was not necessarily a consequence of there

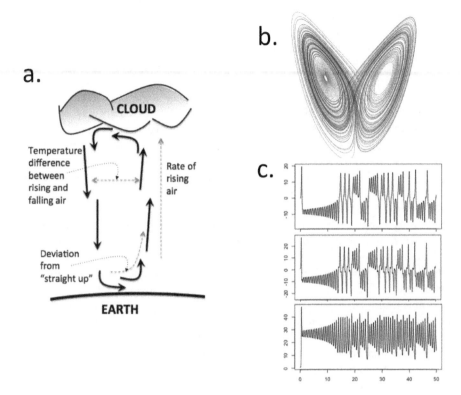

Figure 4.3 The modeling of a convection roll. a. A cartoon version of an idealized convection roll illustrating the three variables used in Lorenz's model (approximately). b. A plot of all three variables in three dimensional space, illustrating the famous "butterfly attractor" or Lorenz attractor, a chaotic attractor (source: Dschwen: https://commons.wikimedia.org/wiki/File:Lorenz_attractor.svg). c. A time series of the same three variables, illustrating the unpredictable nature of the trajectories.

being many factors involved, but rather even an extremely simplified model of just one component of the system led to chaos. Weather systems are unpredictable at their core. The idea of a convection roll and the results that Lorenz found are illustrated in Figure 4.3.

Intuition of the importance of chaos in a simple system

Chaos is well-known in theory. It is also very well-known in a variety of practical applications, such as the pumping action of the heart and weather prediction. Ecological systems are almost certainly examples of chaotic systems, at least some of them, sometimes. Yet our understanding of chaos at the theoretical level dramatically outstrips our ability to use it in practical ecological settings. It is, for example, very likely that biological control systems are chaotic, but gaining

evidence that this is true in the field has proved to be very difficult. Our inten-
tion here is not to summarize the immensely complicated statistical procedures
that people are using today to study the phenomenon in the field. Nevertheless,
there are a couple of technical details that would be useful to understand before
a more general discussion of the meaning of chaos for agroecology.

Most of the insights in ecology from chaos come from attempts to model
populations, with the most evident application to biological pest control. Such
an emphasis has been only for convenience, and it is generally true that any
aspect of the agroecosystem, from soil microbes to farm landscape planning,
might be approached with the insights from modern chaos theory. In particular,
the classic example of chaos is the model we used to generate the example in
Figure 4.2. It is referred to as the logistic map, and envisions an understanding
of populations changing through time in a very seasonal environment, such that
the density of the population this summer is at least partially predictable from
its density last summer – in other words, we look at the population just once a
year. We thus conceive of the population as existing at discrete points in time and
ignore everything that happens between those points (for a review of population
dynamics, see Chapter 2). In general we say that "the number of individuals next
year" equals "a function of the number of individuals now." If that "function" is
a simple multiple, we say the population is changing at an exponential rate. Note
here that a linear function ($y = a + bx$, where $a = 0$ and $y = x_{t+1}$) gives rise to an
exponential growth pattern. So, if every year the population is twice the number
it was the year before, we would see something like a projection in time with
yearly counts something like 1, 2, 4, 8, 16, 32 and so forth, and the population
would simply continue going up at an exponential rate forever (see for example
Figure 2.11). But even a small bit of realism suggests that no population can
have this kind of growth forever, otherwise organisms on Earth would have run
out of space to live millions of years ago. So there must be some sort of control
on the population. This elementary fact makes all the difference in the world.
That "function of the number of individuals now" cannot be linear, which is to
say, the simple statement that the number now is equal to some multiple of the
number last year, cannot be true. It may be true for some limited time frame,
but when the number of individuals in the population becomes too large, it will
be impossible for those individuals to reproduce because they will start running
out of food and the population will simply stop growing. So that "multiple"
of the previous year's population depends on (or is itself a multiple of) the size
of the population. If the number this year is some multiple of the number last
year, we have a linear relationship. But if the number this year is some multiple,
which itself is some multiple, of the number last year, we have a "non-linear"
relationship (see for example Figure 2.11a). It is this essential non-linearity
(further discussed in Chapter 8) that creates the conditions for chaos to emerge.

This example of populations jumping from one year to the next (or one day
to the next, or one moment to the next) is clearly a limited way of looking at
populations. It is effectively assuming that time can be completely understood
as an integer variable, which is to say, progressing in integer time. So we have

time $= 1, 2, 3, \ldots$, but never $1, 1.5, 2, 2.5, \ldots$, and especially never truly continuous. To some extent this is a philosophical issue, reflected in the long history of transfinite numbers (different kinds of infinities)[1] and infinitesimals.[2] But for our purposes here, it is only a practical issue. If we conceive of a population as discrete in time (time jumping ahead in integers), then chaos emerges naturally from the simple assumption of nonlinearity. However, if we conceive of a population moving in continuous time (effectively, if we use a differential equation to describe it), chaos is NOT possible in either a one-dimensional or a two-dimensional form (that is, for either a single population or for two interacting populations), but can emerge only if there are three or more dimensions. This is one of the reasons why the subject of "multidimensionality" is so important, as we discuss in Chapter 7.

A subject that emerges naturally from such considerations, yet receives little attention in the ecological literature, is the question of whether the objects of study are themselves integers. As just discussed, time is frequently thought of as either integer or continuous, depending on the context in which it is being used. But the objects of study (e.g., populations) are most often studied as continuous variables, which obviously makes sense if we conceive of it as some measure like, for example, biomass. But even if we conceive of the population as numbers of individuals, most of ecology treats the population as a continuous variable, even when formulating models of it in discrete time. This is a subject to which we return in Chapter 5, but for now we reflect on its significance with a fanciful example of how a farmer might view a population differently than an ecologist.

Farmer math and the intuition of chaos

When Alexander von Humboldt traveled the world, he made observations as did many travelers both before and after. But he also measured things. Obsessed with accurate measurement and precise description, he was more than only a stereotypic European traveler/scientist. While historians of science mainly laud his careful precision, perhaps it is as legitimate to ask what was the purpose of such precision.[3] Most contemporary analysts would admit that measuring the length of a whale to the nearest angstrom would be pointless. Yet European exploration was as much about conquest as about gaining knowledge, and precisely measuring something still carries with it a, perhaps subconscious, notion of control over that thing. From a logical and scientific perspective, such an attitude is absurd. But sociopolitical influences still abound in science, and unthinking and pointless precision remain well-accepted shibboleths among scientists. Humboldt's personal politics aside, he was a man of the times, and the times suggested an important role for measurement, as precise as possible, in the construction of knowledge.

Von Humboldt's travels were almost a century before the influential travels of Albert Howard, at a time when, although von Humboldt himself was rabidly anti-imperialist, as a practicing scientist of the time he focused his work on careful measurement, clearly standard practice. Contrarily, as we noted in Chapter 1,

Howard was sent to India to "teach" Indian farmers the modern techniques of farming. But Howard made the mistake of listening to the Indian farmers, as we detailed in Chapter 1. Many anthropologists would argue, and we would agree, that listening to local people talk about what they do generally gives a richer understanding than direct measurement of their activities. Not only that, but also it provides inklings of why it is they do it, sometimes providing further grist for the mill of the observer. When talking to poor farmers in Latin America, we are frequently struck with their uncanny ability to understand things that, for us, seem to require more scientific background than they formally have. And it does cause us to wonder if changing course on some well-worn ruts in the bible of knowledge (i.e., the peer-reviewed scientific literature) might lead to different and interesting places.

One general type of farmer's knowledge is clearly mathematical, even though it may not seem so at first glance. For example, the math of the pests on their crops is of necessity a central concern. However, there is a strong tendency in the Western scientific tradition to try and understand the threat from pests by what might be called extreme quantification, as in the tradition of von Humboldt and other European scientists of the eighteenth century. One of the most obvious things to measure is the size of the population of the pest (or a surrogate measure thereof, such as amount of damage to crops). Whether trying to determine current needs for pesticides or predict future potential for biological control, a base-line variable seems to begin with the measurement of the population density of the pest.

Yet the average farmer only partially connects with that idea, sometimes forced to think that way by leading questions. Rather than a base-line density, farmers frequently have their own mathematical framework, seemingly more qualitative, but philosophically perfectly quantitative. At its most elementary, farmers seem to recognize four levels of population: (1) hardly any, (2) some, (3) many, and (4) too many. Almost always there is a further analysis of how those levels are related to one another over time, an explicit or implicit statement about population dynamics. If during the current season there are few pests, what does that say about next season? Perhaps nothing, but perhaps an expectation that there will be more next year, and that more would be worrisome, implying "many" the year later. Whatever the specifics, it is usually the case that the initial four-way quantification (hardly any, some, many, too many) is accompanied by a dynamic rule, or set of rules.

From "hardly any" this season, we might expect either hardly any or some next season (or week or month). From "some" this season, we might expect either "hardly any", or "some", or "many" next season and so forth. This "farmer math" is illustrated in Figure 4.4. The logic is clear. There are evident transitions that are expected to occur from year to year. However, these transitions (e.g., from hardly any to some; from some to many) are not fixed, but rather are alternatives (e.g., some may go to hardly any, or many or stay some). The most obvious way of evaluating these alternatives, one which both farmer and observant agroecological researcher is likely to employ, is to suggest that there

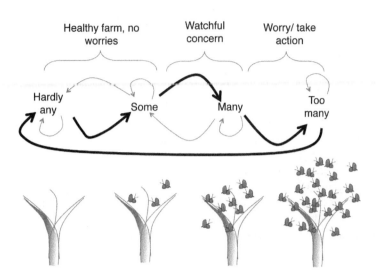

Figure 4.4 Farmer's math regarding pests

are probabilities associated with each transition. This is convenient and works well, to be sure. However, the assignment of probabilities actually carries with it a strong assumption about how nature works – it has a strong random component to it. So, we might say that the fraction of the times that the population will go from some to many is 0.5, and the fraction it will actually decline to hardly any is 0.25, and the fraction it will stay as some is 0.25. This may describe the long-term population behavior adequately. But here we are concerned with ulti-mate meaning. And the process of assigning probabilities suggests a rule about nature, that nature is, to some extent, governed by random processes (later we discuss this more fully within the framework of stochastic processes, the subject of Chapter 5).

There is an alternative. In a sense chaos is precisely the same as randomness, which is to say, the idea that there is a probability (the fraction of time) of going from one state to another. However, in chaos there is no *inherent* randomness in the system. To take a concrete, albeit completely artificial example, suppose when we query the farmer closely we find that her quantitative assessments are based on 10 years of experience of walking around her farm, assessing the number of pests. From the information she provides, we take the von Humboldt approach and assign a numerical value to her observations. Thus we translate too many to be 300, many to be 140, some to be 40, and hardly any to be 10. And her qualita-tive rule is that the population tends to go from hardly any in any given year to some in the next year, to many the next year, to too many the next year, followed by a collapse to hardly any again. Although her detailed knowledge includes the possibility that this projected sequence can be interrupted by years where the category "some" is repeated, or even that the population can revert to a previous stage sometimes, it is thought that on average, the sequence is normally from 10 to 40 to 140 to 300 and back to 10, that is, the bold black arrows in Figure 4.4.

To put these considerations into a formal mathematical form, we can fit a simple mathematical model that includes a term that accounts for reproduction of the pest, plus another term that accounts for how the individual pest organisms interfere with one another when they become too common. We did just that (see the caption to Figure 4.5) and obtained a model that fit the initial data 99.9999% perfectly (in statistical terms the R-square is not distinguishable from 1.00, or 100% goodness of fit of the model to the initial data). In Figure 4.5 we see that running the model predicts pretty well what the farmer expects over a 15-year period, suggesting that the model represents a certain sort of understanding of the underlying ecology of the system. We rightly congratulate ourselves on a very accurate model. However, if we project precisely the same model for 100 years, we see something completely different, as shown in Figure 4.6. The model generates chaotic trajectories!

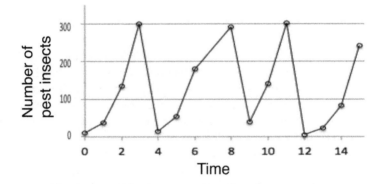

Figure 4.5 Projection of the population over a 15-year time period. Note that the qualitative pattern is much like the farmer's qualitative prediction (*almost none* gives way to *some* gives way to *many* gives way to *too many*). The actual model is: Number of pest insects next year = 3.9988(Number of pest insects this year) −0.01322(Number of pest insects this year)^2.

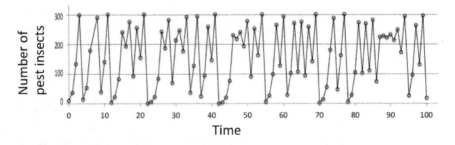

Figure 4.6 The model of Figure 4.5 extended over a 100-year time period

We thus see that the apparent fit of a model to a data set did not really offer any deep insight about the process involved – indeed reliance on the fit of the model to the data, even if the model fit the known data 100% perfectly, could have led to a false sense of understanding. The message here is very clear. Ecological models, even if they correspond almost perfectly with known data, may give completely erroneous results over the long run. This is an important message of chaos, and is a direct result of that basic idea of sensitive dependence on initial conditions.

With reference to the continual real problem of weather prediction alluded to earlier in this chapter, this is also why, even though we are pretty good at prediction of the weather for 5 or 6 days into the future, 5 or 6 weeks into the future remains almost impossible. The weather system is chaotic, just like many ecological systems. Furthermore, the initial ideas of transitions among states (all the arrows in Figure 4.4) with probabilities fixed to each of them, would likely have represented a far better sense of understanding. This would have been a stochastic approach to the process. Indeed, the interweaving of chaos and stochasticity is an active area of ecological research, something we deal with more fully in Chapter 6.

Chaotic attractors, transients and Cantor sets

Recalling the material in Chapter 2 concerning the idea of stability, it is generally thought that a system is either stable or unstable (e.g., see Figures 2.13 and 2.15). Sometimes it may be an oscillatory system, but it is usually thought that some sort of periodic behavior will characterize a system such that in the limit it will approach a forever repeating cycle, a limit cycle. When a system is NOT at a limit point or in a limit cycle it is said to be "transient." Trajectories that are on their way toward either a limit point or a limit cycle are said to be transients.

A general terminology has evolved concerning these categories of dynamical systems. Limit points can be either stable or unstable (see Chapter 2), and limit cycles similarly can be stable or unstable. Recent literature refers to stable points/cycles as "attractors" and unstable ones as "repellors," for obvious reasons. And this leads to possible confusion. A chaotic attractor is, as might be guessed, an object that is eventually approached, and trajectories that are not on the chaotic attractor are transients to it. The idea of an "attractor" that is "chaotic" may seem strange when first encountered. Indeed, that is what everyone thought when we first discovered the phenomenon, which is why another word for a chaotic attractor is a "strange attractor."

Consider the example of an artificial population living on a pair of islands. On each island the population undergoes its normal course of living, with population dynamics similar to that shown in Figure 4.2c (where both populations are plotted in relative terms, ranging from 0 to 1 – with 1 being the maximum population size). However, at each time step some individuals migrate from island to island.[4] If we plot the proportion of individuals (from the maximum) on each of the islands at each time step, we get the strange patterns shown in

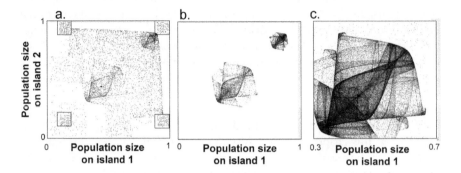

Figure 4.7 Chaotic (strange) attractor of a population on two islands, with migration between them (on each island, the population sizes are scaled to be between 0 and 1.0, the fraction of the maximum population). a. 1,200 distinct simulations, each of which starts randomly within one of the four small squares indicated, and runs for 20 time units. Note the scattering of points (which has a certain pattern to it; i.e., it is not random), but also the especially dense concentration of points in two zones, one toward the center of the graph, one in the upper right-hand corner. b. A single simulation of the same two populations, but with the first 100 time steps eliminated, and the simulation running for 100,000 time steps. Note how the two concentrations of points observed in (a) become more dense and that ALL possible points fall within those two regions – the two regions together are a strange attractor. c. A magnified section of (b) illustrating the complex detail within the strange attractor.

Figure 4.7. In Figure 4.7a we present the results of 1,200 computer simulations. In each simulation, the system starts at a random point within the range of one of the four small squares illustrated, and then runs for 20 iterations (time is arbitrary in this theoretical example – it could be 20 days, 20 weeks, 20 centuries, etc.). The resultant pattern (Figure 4.7a) shows a scattering of points with a certain pattern. Of course there is a large number of points within the four small squares (that is where we began the simulations, and we did 300 simulations starting within each of those squares). But note that there is a pattern at a larger scale also (still referring to Figure 4.7a). The rest of the points are scattered around the whole phase space (but they are not scattered completely at random – we return to this point later). Also note that there are two concentrations of points, one toward the center of the graph and the other in the upper right-hand side of the graph. These concentrations are the points that emerge after the system runs for about 5–10 iterations (time units). Remarkably, if you let the simulation run for a very long time, those two clusters of points remain forever after. So, those two regions are regions that will "attract" all points, no matter where the system is initiated, as shown more clearly in Figure 4.7b (where we ran the simulation for 100,000 iterations, but discarded the first 100 iterations before plotting anything). The amazing thing is that within those two areas (the lower one of which is plotted again in Figure 4.7c), there is a remarkably complex structure. Is it not strange that all starting points should arrive at this attractor? It is thus sometimes called a "strange attractor," a region that attracts all starting points, but itself has a complex internal structure.

As we noted earlier, the categorization of "stabilities" in contemporary understanding of dynamical systems (refer to Chapter 2 if necessary) includes (1) point attractors (stable equilibrium points), (2) point repellors (unstable equilibrium points or cycles), (3) limit cycle attractors (stable limit cycles), (4) limit cycle repellors (unstable limit cycles), (5) chaotic attractors and (6) chaotic repellors. The last category, chaotic repellor, is normally ignored (perhaps unwisely), since it would be, in principle, quite difficult to actually observe (the trajectory of the variable would be apparently chaotic, but would be moving away from the chaotic structure itself, but not necessarily away from the qualitative concentration of points within the chaotic attractor).

Dynamical systems that are not at (or near) one of these categories are simply referred to as transients. And here the subject of chaos can become confusing. Although formally unpredictable, there are clear patterns that are frequently emergent from chaotic systems, as can be clearly seen in the Lorenz butterfly chaotic attractor (Figure 4.3b), or the two-island attractor (Figure 4.7). There is a surprising, almost illogical, feature of chaotic attractors that is difficult to understand completely. It is effectively a philosophical point rather than a practical one, but impinges on the interpretation of other aspects of complexity. If a system is truly chaotic, no particular state can ever occur more than once. So, for example, in the two-island case (Figure 4.7), the chaotic attractor is an oddly shaped cluster of points. And, as we noted, there are obvious states that the system will never attain (all the areas outside of the main cluster of points), and those points that were initially outside the main clusters (Figure 4.7a) are logically called "transients." However, somewhat surprisingly, no matter how long the model system runs, no particular point will ever be repeated. This means that for a chaotic system (1) we can predict that the system will be within a particular region (e.g., inside the butterfly shade in the Lorenz attractor, or within one of the two main clusters in the two island attractor) but (2) we cannot predict the precise location of a point. We can run the model till the end of the universe, and no point will ever be repeated (of course, this is a bit theoretical, both because of our anticipated end of the universe, and because the numbers we are dealing with are real numbers, which is to say, not integers). So, for example, one of the points in Figure 4.7 is $x = 0.50012$ and $y = 0.51137$. That point will never be repeated no matter how long we run the model. It may seem sort of "angels on the head of a pin" to ask whether that $x = 0.50012$ and $y = 0.51137$ indeed could never crop up sometime in the future, which is why we say it is sort of a philosophical point. However, within that philosophical framework it seems a bit problematical that, in principle, a point will NEVER be repeated. How can we have well-defined finite areas within which points must be located (as in Figure 4.7b), yet those points are completely unique? This might not seem too difficult if we just imagine that eventually the entire sub-region will just be black with very dense coverage and we can have an infinite number of points. That is fine, as far as it goes, but chaotic attractors very typically are not "dense". They typically have obvious "pattern" to them (as in Figure 4.7c). Furthermore, that pattern may be in effect repeated at infinitely many scales, which is to say it is fractal, such that if you take subsets of points, they appear to have a pattern, and this repeats itself, no matter how finely you sample. In Figure 4.8 we see that subsampling the two-island attractor reveals further non-random structure, which may not have been obvious at the larger scale.

This fractal pattern suggests something quite outlandish. A given fixed space within a chaotic attractor has an infinite number of points, as it must if no particular point can be reached more than once even if the model is run for close to an infinite amount of time. However, that space must also contain an infinite number of subsections with no points at all (in order to have a revealed fractal pattern).

There is something both paradoxical and important here, for which an abstract structure helps us understand. It is a structure that was recognized in the nineteenth century by the great mathematician Georg Cantor, well before anyone understood anything about chaos. However, Cantor dealt with the issue in a formal mathematical sense, generating ideas about infinity that some say drove him insane. Fundamental to these ideas is the basic structure of what is known as a Cantor set, which is effectively what the points in a chaotic attractor (e.g., Figures 4.7 and 4.8) are like. Consider a line. Erase the middle third of the line. This gives you two line segments. Erase the middle third of each of those segments. This gives you four line segments. Continue this process forever. The result is a set of points that is uncountable (the mathematician's word for infinite). But more surprising is that as you continue this process to infinity, the area originally occupied by the line approaches the state of having no points at all. In essence this Cantor set has an infinite number of points, contained in a space

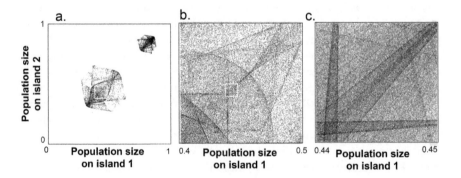

Figure 4.8 The same two-island attractor of Figure 4.7. a. The basic attractor (same as Figure 4.8b) with a small square to be subsampled. b. The subsampled square from (a), illustrating newly revealed pattern within the attractor. Also note the small square to be sampled again. c. The subsampled square from (b), again illustrating newly revealed pattern within the attractor.

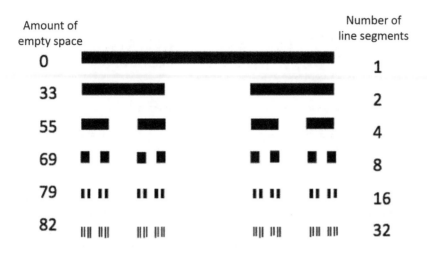

Amount of empty space		Number of line segments
0		1
33		2
55		4
69		8
79		16
82		32

Figure 4.9 Construction of a Cantor set, illustrating how, in the limit, there will be an infinite number of points (line segments), and 100% of the original area will be without points (empty)

that is 100% empty of points! The process is illustrated in Figure 4.9. So we see, logically, how we can have what seems so illogical in a chaotic attractor – an infinite number of points located in a finite space that is empty!

Ecological chaos in the real world?

There are, we believe, compelling reasons to believe that ecological systems are frequently, perhaps almost always, chaotic, or at least have chaotic elements embedded in them. Unambiguous detection of them has proven to be extremely difficult for a variety of technical reasons. Frequently the data we have to work with are insufficiently dense. For example in Figure 4.10 we graph a rare case of a very long time series, deduced from historical records of locust infestations in China, along with a known chaotic trajectory and a random sequence of numbers (before you read the caption see if you can guess which is which). Even here, with the longest known time series of an insect species (as far as we know), it is not completely clear whether we have a chaotic system or not. Yet in many *model* systems, both laboratory experiments and mathematical models, the existence of chaos is clear.

While it is difficult to say for sure whether a given data set is chaotic or not, and many very smart people have been trying to do that for a long time, there are some generalized qualitative structures that strongly suggest chaotic systems, especially when we add socioeconomic factors. Consider, for example, a simple example of a group of farmers making decisions about their farming operations. There is an obvious time lag between making the decision about what to plant

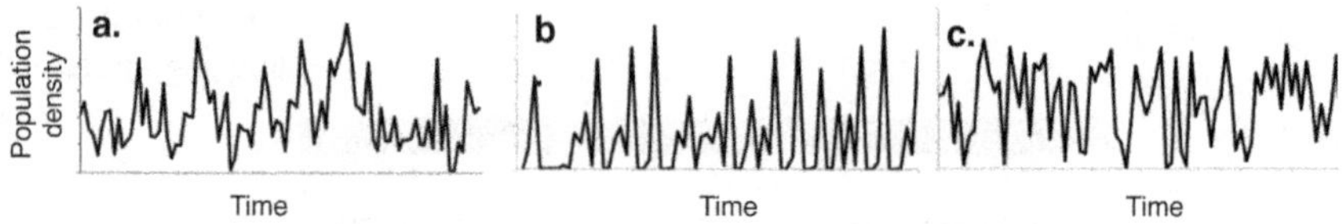

Figure 4.10 Three time series illustrating chaos, randomness and an actual 1,000-year time series of locust populations.[5] a. The actual time series of locust infestations in China. b. A chaotic population in theory. c. A sequence of random numbers.

and the harvest and marketing of the produce, such that the farmers must take the information they have today to try and predict market conditions at harvest time. If all the farmers in a region decide that tomatoes will fetch a high price at the time of harvest and all plant large quantities of tomatoes, as a consequence, there will be a glut of tomatoes at harvest time over the whole region and the price will collapse. But will farmers not realize this and adjust their planting plans accordingly? Perhaps, but herein lies the germ of chaotic behavior.

In a series of interviews one of us conducted with tomato farmers in Costa Rica, we found that, yes, as the current price of tomatoes increases, farmers reported that they would increase the share of their farm planted in tomatoes up to a certain point. But, somewhat unexpectedly, they reported that they would plant less area in tomatoes if the price went above a certain point. The data from these surveys are presented in Figure 4.11.

The basic form of the way farmers make decisions about production, coupled with the elementary notion that a glut will lead to lower prices, strongly suggests a chaotic system. Unable to know with certainty their neighbors' plans (that is the assumption here), they use market prices as a cue, but they know perfectly well (as they said in the interviews) that if everyone plants tomatoes, the market price will collapse. But if everyone pulls out of production, the one who goes full force with production will gain considerably. Much like the famous bar where "no one goes there anymore because it's too crowded,"[6] the tomato farmer must contend with the parallel problem that no one wants to produce tomatoes because too many people are producing tomatoes. This is a structure that very easily leads to chaotic price and production trajectories.[7]

A useful practical example was provided by Japanese agricultural economist Kenshi Sakai and his collaborators with an analysis of piglet production and price trajectories between 1967 and 1991, during which the second oil crisis of 1979 intervened.[8] Through a careful and very complicated analysis they showed that before the oil crisis the trajectories of both production and price appeared to be chaotic, but that after the crisis the overall structure changed to a limit cycle. We present their data for prices in Figure 4.12, along with a regular sine wave fit to the second half of the data set (where the regular oscillations take over from the chaos).

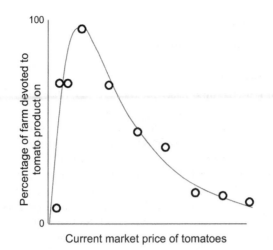

Figure 4.11 Results of interviews with tomato farmers in Costa Rica where they were asked how much tomatoes they would plant in relation to current market prices. The smooth curve was hand-drawn. Based on data from Vandermeer, 1990.

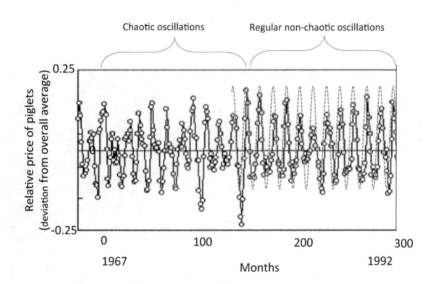

Figure 4.12 Relative prices of piglets in Japan from 1967 to 1992. Note that there was a change from the second oil shock from a chaotic regime to a regular limit cycle. We added a regular sine wave to the right half of the data to make it easier to see that the oscillations are regular (dotted red curve). A glance at the left-hand side of the graph suggests that no regular oscillation curve could be invented that would fit the data very well.

Source: Base figure reprinted with permission from Nonlinear Dynamics, Psychology, and Life Sciences. Sakai et al., 2007.

The generalized structure of chaotic attractors

Although the bulk of popular attention to the phenomenon of chaos has focused on the unpredictable nature of chaotic systems, such a focus is in our view misdirected. Consider, for example, a tornado (Figure 4.13). It is absolutely true that a tornado is a chaotic object, but that fact does not imply that there is no structure to a tornado. The formality of chaos means that if two dust particles in the tornado are very close together, even touching one another, at some particular point in time, that fact provides us with no information at all about where they will be with respect to one another in the next few minutes (as in the case of pouring flour into water in Figure 4.1). Such is the consequence of sensitive dependence on initial conditions. However, we can say with some certainty that both dust particles will remain within the structure we know as a tornado. That is, a tornado is a recognizable object and the fact that it is a chaotic object does not mean that it is somehow out of reach of our understanding. Sensitive dependence on initial conditions is an important concept with regard to our ability to be able to predict the precise condition of the system, but this simply means we must look to other goals in its study.

This idea of trying to understand the "morphology" of a strange attractor is at first simple (e.g., the tornado), yet can take on a more complicated form. Recall the population living on a pair of islands (Figures 4.7 and 4.8). For a wide range of parameter values this system is formally chaotic. However, in that chaos there

Figure 4.13 A tornado touches ground near Anadarko, Oklahoma.

Source: Daphne Zaras [Public domain], https://commons.wikimedia.org/wiki/File:Dszpics1.jpg.

is a pattern, sometimes a complicated and extremely structured pattern (as we noted in Figure 4.7c). A quick glance at Figure 4.7c shows that only a subset of possible points is reachable once the system is in chaos, and within that subset, some regions are more likely to have points than other regions. Even though we cannot predict where any point will be within the area housing the chaotic attractor, there are some regions within that attractor that have a higher probability of containing a point than others. That is "structure," and it is the sort of structure that should attract the attention of research scientists, and maybe even some day will have practical applications for agroecosystems.

To recap, sensitive dependence on initial conditions means that knowledge of the position of a dust particle within a tornado conveys almost no information about where that dust particle will be in the future (if indeed it is on a chaotic attractor). However, the most important thing about this structure is that we CAN say with some certainty that the particle will be within the structure we recognize as a tornado, or, in the example of the two populations on the islands, the more abstract structure of the island populations in Figure 4.7b. We should ask questions, as we do, about how strong the currents within the tornado are, what is its overall form, how fast is it moving and so forth. In this sense all chaotic attractors are similar. Precise prediction is neither possible nor desirable. Understanding general form should be the goal. Yet here we face some clear challenges. The simple question of whether or not a system is chaotic has turned out to be relatively uninteresting, yet the important question of what is the structural nature of chaos has not yet received as much attention as it deserves. A major issue, in our view, is elaborating allowable versus unallowable outcomes. It is a very subtle issue, but could have important practical consequences.

An interesting example of a chaotic structure which serves to examine the structure of chaos more carefully, emerges from some simple considerations of what we might expect from the escape of GMO transgenes in plants modified to contain the toxin of *Bacillis thurengensis* (Bt), thus rendering them resistant to attacks from the main caterpillar pests.[9] The question naturally arises, what will be the consequences of the escape of this gene (i.e., the escape out of agricultural planning of individual plants that have the gene in them) into the environment, something known to occur through cross-pollination of related plant species. This framework presumes that the non-modified plant will be competitively superior to the modified one (based on the assumption that some aspect of fitness will have been sacrificed in the original genetic modification – a common assumption among transgene developers), but that having a certain number of transgene individuals in the population will act as an effective barrier to the spread of the pest from one plant to another (metaphorically similar to the herd immunity of epidemiology, where the presence of resistant individuals in a population reduces generalized transmission of a disease). In essence the overall dynamic structure of this situation is summarized in Figure 4.14.

The general dynamics of this deceptively simple case are very complicated, but for our purposes here, for some parameter values it forms a chaotic attractor with a very clear form, as shown in Figure 4.15. In this model, a general structure is

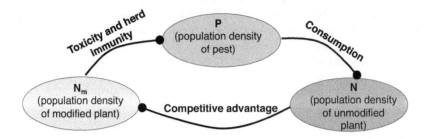

Figure 4.14 Theoretical structure of an escaped transgene conferring resistance to attack from caterpillar pests through the Bt toxin[9]

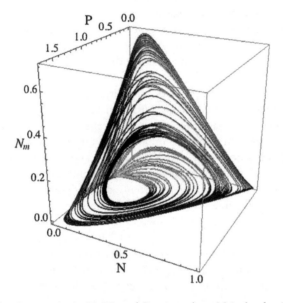

Figure 4.15 Chaotic attractor in N, N_m and P space, where N is the density of the non-genetically modified plant, N_m of the genetically modified plant and P of the pest

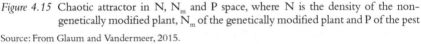

Source: From Glaum and Vandermeer, 2015.

easily visible, in which oscillations between the normal plant and modified plant cycle around when the pest is at a very low level, but the pest then oscillates with both plant populations at other times. It is a chaotic attractor, so a more detailed description of its behavior involves some new concepts (which we introduce later), but its overall form is fairly obvious.

To understand some underlying features of this system we can examine a similarly structured but simpler chaotic attractor, the Rossler attractor, which is famous among mathematicians for its utility in understanding chaos. It is well known that in a continuous system, only three dimensions and above can yield chaos, and perfectly linear systems cannot yield chaos. The Rossler attractor is interesting for theoretical reasons in that it is probably the simplest mathematical form that generates chaos, having only one nonlinear term in an otherwise completely linear system. We illustrate the Rossler attractor in Figure 4.16, for three arbitrary variables, where it is obvious that its structure is similar to the attractor inherent in the problem of an escaped Bt transgenic plant (compare Figure 4.15 with Figure 4.16).

Contemplating the Rossler attractor (or equivalently the Glaum attractor in Figure 4.15), we can get a sense of its underlying structure, and indeed, the important underlying structure of chaos itself. Note that the vertical loop seems to stretch up and then come down into the flat circular pattern (as we follow the vectors from position 1 to 2 to 3 to 4, in Figure 4.16). Note how the attractor seems to actually fold onto or into itself – vector number 3 actually folds itself into the horizontal circular pattern. It is this folding and stretching of a vector field that is characteristic of chaos. We further diagram this process in Figure 4.17.

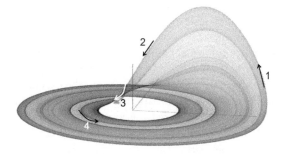

Figure 4.16 The Rossler attractor (the three dimensions are completely arbitrary)

Source: By Wofl, https://commons.wikimedia.org/wiki/File:Roessler_attractor.png.

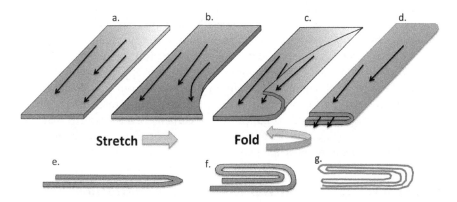

Figure 4.17 The basic modification of the vector field to create a chaotic attractor. a. Begin with a normal (unremarkable) vector field. b. Stretch the end of the vector field out. c. Fold the stretched out part over. d. Complete the fold. e. Now, just looking at the cross section of the original vector field, stretch out the folded field. f. Fold over the stretched out field again. g. Stretch and fold again. The whole procedure goes from one layer (a) to two layers (d) to four layers (f) to eight layers (g). Continue this process forever, and you get the basic structure of the vector field of a chaotic attractor.

With the process of stretching and folding (Figure 4.17) we see that the structure of the Rossler attractor (and the Glaum attractor) is actually an infinity of layers, and we see how it is possible that an infinity of trajectories can coexist without ever crossing one another and without ever repeating a single value. Each time around the trajectory of the attractor is effectively on a different layer, even though it appears to be all in the same plane. And furthermore, we hope it is clear how a cross section through such an attractor gives us points that are infinite in number, but with empty space that spans the entire attractor, the characteristic of a Cantor set (Figure 4.9). The folding and stretching process can also help us understand how two points can start by being very close to each other but quickly diverge in unpredictable ways if we visualize them as starting off at what seems to be almost the same point on the attractor, but effectively on distinct layers (Figure 4.17).

Given this underlying structure we are faced with something of a conundrum. On the one hand, we are cautioned to be careful about complicated computer models that may seem to function well, which is to say, may seem to predict nature over the short term, yet long-term prediction runs up against the problem of sensitive dependence on initial conditions (which we now see can be visualized as two points starting off at what seem to be the same point on the attractor, but effectively on distinct layers). Those models may be excellent tools for short-term predictions, but, much as predicting the weather, we should exert caution when trying to make predictions over the long haul. But the other side of the conundrum is that chaos does have structure, indeed a great deal of structure. In trying to understand those parts of nature that are effectively chaotic, we need to step back and ask questions about larger structures, ask about the morphology of the tornado rather than the position of dust particles within it.

Finally, it is worth noting that the subject of chaos is at the nexus of several subjects normally thought of in the context of ecological complexity. For example, the coupling of oscillators, as treated in Chapter 5, is a classical subject that produces complex patterns of chaos and that was only understood vaguely before our more recent understanding of chaos generally. Likewise, the interplay of chaos with stochasticity has brought us to the brink of application of ecological complexity to real field data, in addition to enriching our understanding of the stochastic nature of ecosystems. Finally, the spatial dynamics of ecological systems (e.g., the Turing effect) has produced certain chaotic patterns that not only reflect complex trajectories in time, but also seemingly random spatial patterns. Many of these topics will emerge again in subsequent chapters.

From cooked carrots to chaos to attractor reconstruction

During some period of time in the 1960s at the University of Michigan, all incoming students were given a multiple choice test with seemingly random questions. Since one of the questions was "do you like cooked carrots or raw carrots better?" the test became locally known as the cooked carrots test. The idea was that if a group of first-year students were given this test and their academic path were followed in detail for the next four years, there would be a certain amount of predictability generated by correlating the scores on the cooked carrots test with the students' academic choices during the subsequent four years. So, for example, a student who preferred cooked carrots, went to bed early, liked action movies, listened to classical music, and hundreds of other preferences or acts, could be counseled that others who shared those answers were highly successful if they became math majors, but would fail if they went into world history. The test was never very functional, perhaps for obvious reasons, but we bring it up here for two reasons. First, it represents a philosophical position that we strongly disagree with, and second, it has reemerged in a slightly different form, a new application of theoretical principles typically associated with chaos theory.

As a philosophical principle, the test was self-identified as completely "nonscientific," aiming only at prediction, not understanding. Whether one cites Popper, as we did in Chapter 1, or any historical framing of either natural or social scientific ideas, the idea of predicting something without understanding why is anathema to science in its broadest sense. No one ever suggested that there was any deep reason to expect that your preference for cooked carrots provided some mechanistic understanding of why you might be good at math. It was correlation without understanding, even suggesting that the fundamental goal of intellectual activity should be prediction alone, with understanding at a deeper level somehow old-fashioned.

The fundamental idea has re-emerged within the general framework of chaos theory. A couple of decades ago ecologists became fascinated with a major theoretical breakthrough in the mathematics of chaos, the Takens theorem.[10] Basically the framework behind the theorem is identical to the cooked carrots test, but with more sophisticated statistical techniques and computer technology. The idea is especially attractive in a system that is chaotic, or at least close to being chaotic. It is a method that takes all known data about a phenomenon and "reconstructs" what is most likely to be the generalized chaotic attractor and then interrogates that attractor. It is actually quite elegant in its analytical framing, but in the end effectively says we should eschew the traditional intellectual goal of understanding a phenomenon and simply try and predict it.

A simple example shows how this project works, in theory. Consider the data shown in Figure 4.6 and extend that model for 500 years, as we have done in Figure 4.18. Interestingly, if this data set were completely random, prediction of the future would be, in principle, impossible. However, if there is underlying structure, even if perfectly chaotic, a certain degree of prediction is possible, indeed very accurate prediction if the data are dense enough. Suppose that the "present" is defined by time = 400 in Figure 4.18. If we have all the data from previous years available, one thing we can do is look for a moment in the past that corresponds as closely as possible to the present moment. In this example, the population density at the present time (time = 400, in Figure 4.18), is approximately 41.6. If we go back in time we find a similar number (41.9) occurred at time 148. We can use the data from starting at time 148 to act as a prediction of

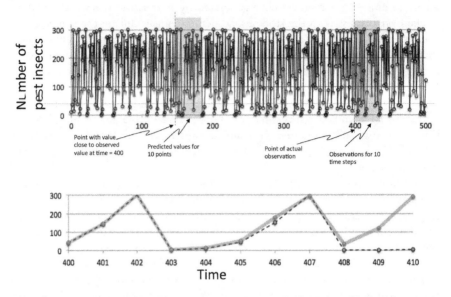

Figure 4.18 Essentials of "attractor reconstruction" (using the formality of Taken's theorem).
a. Assuming we have 500 years' worth of accurate data on the number of pest insects and locating a population density that closely approximates the population density at time 400 (which we assume is the present time). b. Prediction (transparent solid line) and observed (black dashed line) points for the next 10 years, illustrating the relative precision of the reconstruction for at least a short period of time.

what will happen as we go forward from time 400. In Figure 4.18b we see both the predicted numbers (the weighted transparent line) and the actual number (the black dashed line). It is actually rather remarkable that such an excellent prediction can be made. However, the practicality of this methodology is somewhat limited by the reality that hardly ever do we have the kind of dense data necessary to put the method into practice.

What we present here is a characterization of the procedure based on the theorem of Takens. In reality there are many important and innovative additions to the methodology, making the reality of predictions somewhat more confidence inspiring, but beyond the intent of this text. Here we want to acknowledge this technique (formally called attractor reconstruction) and its tremendous potential. But we also highlight some technical problems that need attention. For example, if the attractor is chaotic but very complicated, the amount of data needed for a prediction could be extremely prohibitive. The classic metaphor of the black swan is sometimes the most important story, and this method strongly presumes all swans are white.[11] Or, if there are alternative chaotic attractors that interpenetrate one another, a very small stochastic force, or, for that matter, a small change in the value of one of the parameters, may push the system into a completely different ecological regime. These are issues that will be covered in more detail in later chapters. We mention them here only as cautionary for the methodology suggested.

However, we do wish to emphasize the important philosophical problem we mentioned earlier – the attractor reconstruction procedure is basically a cooked carrots type of procedure. There is no pretense of trying to understand the system under consideration, only predicting where it will go, which is sometimes all we want and certainly is the basis of the entire field of engineering. But, if *understanding* is the true goal of science, whether the natural or social kind, attractor reconstruction is simply not science.

Conclusions

The subject of chaos is now a standard subject in many fields, and certainly central to the idea of complex systems. Much of the current literature is difficult to wade through due to its heavy mathematical content, but the basic ideas, which we have attempted to present here, are fairly clear. Particular applications tend to be quite dense. An excellent series of papers by Kenshi Sakai and his group[12] is well worth consulting for the reader interested in more advanced applications to agriculture, especially his 2001 book.[13]

For our purposes in this book we have attempted to give the reader a qualitative feeling for the subjects that chaos stimulates. The very idea of sensitive dependence on initial conditions is crucial to the idea of chaos, but not really the central message. At a very theoretical level, almost a philosophical level, ideas such as Cantor sets and repeating fractal patterns remain key elements of the chaos literature, even though their practical application is thus far elusive. At a more qualitative level, the idea of attractor shape or morphology is a subject deserving considerable attention for those interested in agroecology. If production systems are chaotic (which we believe they frequently are), what is the overall qualitative structure of that chaos, and will an understanding of it help us design and manage the agroecosystem better than we do now? Finally, it is worth noting that, as will be evident in later chapters, the subject of chaos intersects with various other aspects of ecological complexity, meaning that understanding its basic features is ultimately important for understanding our vision of ecological complexity.

Notes

1 For an entertaining romp through this subject see Wallace, 2010.
2 Infinitesmals are the foundation of calculus, which the Jesuits of the sixteenth century found challenging to say the least; see Alexander, 2015.
3 Cannon, 1978, chap. 4; Dettelbach, 1999.
4 Vandermeer and Kaufman, 1998.
5 Stige et al., 2007.
6 This quote is frequently attributed to the famous baseball player Yogi Berra.

 7 Vandermeer, 1990.
 8 Sakai et al., 2007.
 9 Glaum et al., 2012; Glaum and Vandermeer, 2015.
10 Takens, 1981; Huke, 2006.
11 This is a metaphor used to denote a very rare event that is usually a surprise when it occurs. It is based on the fact that swans in Europe always are white and it was natural to conclude that it was a general rule of nature that swans are white. But then, in 1697 Dutch explorers in Australia came upon a swan with all black feathers, falsifying the generalization that all swans are white with a single observation.
12 Ye et al., 2007; Sakai et al., 2008.
13 Sakai, 2001.

5   Stochasticity

Introduction: deterministic versus stochastic

As you decrease the temperature, a pan of water eventually turns into a pan of ice (at 0°C). It is certainly reasonable to simply state that the transformation of the water into ice is "determined" by the temperature. Similarly our feeling of comfort when we walk around outside is at least in part "determined" by the temperature, among other things. The world is generally thought to be "deterministic" in this sense. One thing determines another. Mathematically this is usually represented as a function, such as $y = f(x)$, for which we say y equals a function of x, which is really quite the same thing as saying y is determined by x. In some ways it is the underlying foundation of science to have things determined by other things: we live in a deterministic world.

Yet in actual experience, things like our feeling of comfort when we walk around outside is not determined by only temperature. Many and varied things determine it, each in part. The humidity, whether it is raining, with whom we are walking, the slope of the land, the amount of ice or mud or sand or, for that matter, ball bearings. A large number of things could be involved in that feeling of comfort, and we could probably just sit down and list them, at least develop a very large list of things that might be involved in "determining" our comfort. Yet a moment's reflection suggests that it would be kind of difficult to list ALL of them. If we tried, we would likely come up with many things, but would we ever say that we now have a list that determines 100% of our comfort? And would there not be some residual effect in the sense that today we felt a bit more comfortable than yesterday, even if all those things we listed are precisely the same today as yesterday? The realization that we will never be able to list all the determining factors suggests that there is something more than determinism involved, at least in practice.

That is basically the operational sense of "stochasticity." There are factors that influence our comfort that, for all practical purposes, we will never be able to know. Yet we know they are there. They are the random or stochastic factors that are part of the process that "creates results," the results come from the things that determine them plus the random factors. The world of deterministic science views its fundamental goal as figuring out the factors that determine an observation – basically assuming that the world is deterministic. Yet a full accounting of the observation is likely to include things that we do not know. We conveniently treat them as random or stochastic. We then divide the world into deterministic and stochastic.

It is unlikely that anyone would deny that random perturbations, in other words, stochasticity, influence ecological processes, or at least the way we can measure and understand them. Stochasticity is taken as a given in most ecological framings of models, theories and experiments, and it has become a standard component of the ecologist's toolkit in the form of both classical and modern statistical methodologies. However, formerly, the operation of stochastic forces was thought of as something to be minimized, something that generated an annoying "fuzz" around the "real" value of whatever was being measured. Experiments and observations were then organized around the goal of minimizing this "noise" so as to get at the underlying truth. While this is a sensible position in many contexts, it is not so within the framework of ecological complexity. Indeed, stochasticity takes on a completely different meaning – it can be the main determinant of the qualitative behavior of a system. It can be the key to understanding, rather than a pesky artifact that annoyingly interferes with understanding.

The importance of stochasticity

The Newtonian Revolution ushered in over a century of confidence. The predictability of planetary motion resulted from the same equations that tell where a cannonball will fall, to say nothing of the rate of fall of an apple. That confidence was broken when we tried to apply the same ideas to the structure of the atom. Neils Bohr proposed that the electrons were like small planets orbiting a nucleus that behaved much like the sun, gravitationally attracting those tiny planets (electrons). While the details are not particularly relevant to our story here, the Newtonian gravity model just did not work for these very tiny "particles." A close look at those electrons revealed such strange properties that even today we remain somewhat dumfounded, and certainly have no sensible metaphors from common experience that help us explain their behavior. However, the conundrum was resolved by a key paradigm shift: application of a stochastic model. That is, the very position of the electron, indeed even the definition of the word "position" came under scrutiny. The crucial insight was envisioning the electron position as a probability. As we move away from the nucleus of an atom, there is a probability that we will encounter the electron, and that probability increases as we move away from the nucleus, to a point at which it begins to decline. Note that this relatively uniformly accepted theory is a very large qualitative deviation from the previously triumphant equations of Newton. No fixed positions at all, just probabilities. The addition of a stochastic framework changed our notion of physical dynamics at the quantum particle level, and the resultant quantum dynamics became the new standard of understanding. Einstein was never comfortable with the new approach and even the famous popularizer of physics, Richard Feynman, thought of the new theory as just a practical calculating device, lacking the deep insight usually provided by theory. Yet what matters here is that adding randomness, a stochastic component, changed our view of the world qualitatively. Such a revolution may also be emerging in ecology.

An agroecological example may help at this point. In our ongoing research in Mexican coffee agroecosystems, the ant *Azteca sericeasur* figures as a major keystone species.[1] In particular, it is a major predator where it forages, especially notable for its potential effect on the coffee berry borer, one of the most significant pests of coffee worldwide. The ant nests in shade trees and forages on the nearby coffee bushes. Observations on the foraging activities of the ants reveal that they do not maintain activity at a constant level, but rather are sometimes very active and sometimes not so active, partially depending on how far the coffee bush is from the nesting tree. The coffee berry borer needs approximately one hour to burrow into the coffee fruit, during which time it is poised with its rear end sticking out of the hole it is boring, presenting an easy target for any ant that finds it (Figure 5.1). Thus, for successful entry to the seed, the beetle needs an hour of ant-free time, but once inside of the seed, the beetle is completely protected from the predaceous activities of the ants.

Since the ants are variable in their foraging activity, exactly how that foraging occurs will determine whether the beetle will be successful or not. For example, if the ant colony has regular activity cycles that are approximately one hour long (one hour of intense activity followed by one hour of very low activity), the beetle can easily find a full hour to bore completely into the seed before it is encountered by an ant. However, if the activity cycles are approximately 20 minutes long (20 minutes of intense activity followed by 20 minutes of very low activity), the beetle will never be able to encounter a full hour of low ant activity to be able to bore into the seed without being first discovered by an

Figure 5.1 The coffee berry borer attacking coffee fruits, and the *Azteca* ant responding by removing the berry borer from the fruit

Source: Authors.

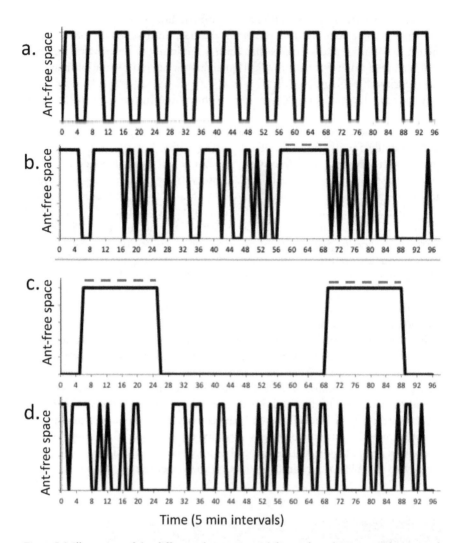

Figure 5.2 Illustration of the difference between variability and stochasticity, with the attack of the coffee berry borer occurring only when there is ant-free space, and only when it lasts for a continuous 60-minute period of time (each tick in time is 5 minutes). a. Regular variability of 20 minutes with ants alternating with 20 minutes without ants, and thus providing no opportunity for the berry borer to enter the fruits. b. A random sequence of the same number of 5-minute intervals of ant activity (and non-activity), illustrating how a stochastic offering of the variability can provide the beetle with an attack opportunity (ant-free period of at least 60 minutes, illustrated by the dashed horizontal line). c. Regular variability of 120 minutes with and 60 minutes without ant activity, providing two windows of opportunity for the beetle (indicated with dashed bold red lines). d. The same number of 5-minute intervals, but offered in a stochastic sequence, showing a lack of any opportunity for the beetle (we chose this particular sequence because it did NOT have an opportunity for the beetle; the sequence in (b), equally likely, would have given one window of opportunity).

ant. In Figure 5.2 we see the difference that stochasticity can make. Regular variability has a very different outcome than random variability (i.e., stochasticity). If, for example, the ant activity cycles occur regularly at intervals of 20 minutes (four 5-minute periods of inactivity alternating with four 5-minute periods of activity) (Figure 5.2a), the beetle will never encounter a period long enough (60 minutes) to be able to bore safely into the seed. But on the other hand, if 5-minute cycles occur at random, there will be times where ant activity will be low for at least an hour and the beetle will be able to bore into the seed and be protected from ant predation (Figure 5.2b). Interestingly, we could also have the reverse situation, where a regular ant activity cycle can provide enough time to the beetle to bore into the seed (Figure 5.2c), but presenting the same number of 5-minute activity periods stochastically will eliminate that opportunity (Figure 5.2d).

Basic population processes with stochasticity

Dealing with stochasticity in ecology is not without historical precedent. An early controversy in population dynamics was ultimately related to the question of stochasticity.[2] It had to do with how populations in nature were "controlled," with one side claiming that there was a more or less continual response of populations to their own density (lower population growth rate at higher population densities), while the other side claimed that populations grew continuously and exponentially until some random environmental factor, frequently thought to be associated with the general term "weather," acted to kill large numbers of individuals in the population, thus reducing it dramatically. This density dependent versus density independent debate was in essence also a debate about the importance of random factors in the environment, with the density independent proponents suggesting that random environmental factors were essential elements in population control.

Reflecting on the earlier argument about density dependence, proponents of the density-dependent school argued persuasively that even if the population growth rate was exceedingly small, but positive, the population would eventually "overpopulate," which is to say, would grow exponentially toward its carrying capacity. There needed to be, according to this argument, a way in which the population could have some sort of feedback (see Chapter 2) such that when it got bigger, its growth rate would be adjusted accordingly. The density dependence proponents scoffed at the idea that random factors could "control" the population. They postulated that even if populations were assigned growth rates at random, those which happened to be assigned negative growth rates would simply go extinct, and be of no further concern, while those assigned a positive growth rate would grow exponentially, until running out of resources, and therefore stop growing. This means that population control will be exerted at high densities, which means that the control is dependent on the density of the population. Their argument was convincing to be sure. However, it was an argument made on the assumption of strict deterministic processes, perhaps adjusted superficially, but fundamentally based on populations either having a positive growth rate and increasing forever or a negative growth rate and thus rapidly dying.

Advocates of the density independent school answered that when stochastic factors are included, they can in the end not only restrict or limit the size of a population, but also determine extinction versus survival. As an example, suppose a group of isolated *milpas* (small-scale corn and bean fields) are colonized by a soil-born fungal pathogen, such that each year it reappears, providing it had enough propagating material in the soil during the non-cropping season. Will that pathogen be a problem? If so, will it persist forever? Will it persist in all the farms? The strict deterministic answer would be that, while occasionally not able to generate a successful population due to perhaps a random set of environmental factors that limit its population growth rate, once established, the fungal pathogen will grow to become a persistent problem. However, if the population growth rate varies, not from population to population (i.e., from farm to farm), but rather from year to year on each of the farms, something different happens. In Figure 5.1 we illustrate 10 computer simulations of a population growing according to a very low population growth rate (imagine these are 10 farms each attacked by the very same soil fungus, for example). The growth rate of each population in this case changes stochastically every year, under the fairly obvious assertion that each year is slightly different from each other year. Some of the populations (Figure 5.3) appear to grow exponentially for a while, but then suddenly decline, one of the populations (the ninth one) goes extinct after about 60 years. While there is a very essential unpredictability about this system, due to the stochastic force that is imposed on it, there is no suggestion that it cannot be "controlled" by the stochastic factor that was added.[3] Here we have an example of a qualitative structure that is determined by the effect of a

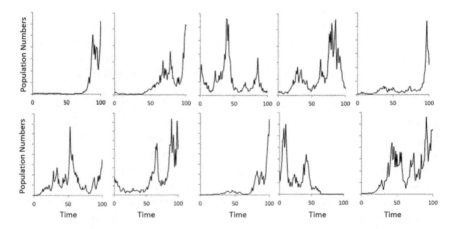

Figure 5.3 Ten populations, with small growth rates, that are adjusted stochastically every year for a period of 100 years. No self-regulatory force is involved, and the "control" of the population is exerted by the stochastic force imposed.

stochastic force, where any attempt to understand it without consideration of that stochastic force, would be folly.

The central role of stochasticity as a force capable of generating qualitative behavior can be demonstrated with other, slightly more complicated, yet still elementary models, an example of which is motivated by that early debate between the density dependence versus density independence proponents. In Figure 5.3 we illustrated how a simple population growth model with no density dependence factor can generate very different population trajectories, all including some level of population reductions (including extinction of the population) just from assuming that the density independent growth rate was stochastically driven each year. Considering the density dependent case, Levins[4] noted that simply adding a random element to the growth rate of a population that is under density dependent control would lead to a change in the qualitative structure of the population. As noted in Chapter 2, a standard mode of population behavior begins with an increase in numbers, but as the size of the population reaches higher levels, there is feedback on the growth rate as the population approaches its carrying capacity and the population eventually stops growing and remains at this equilibrium point. If this population lives in an environment that imposes a strong stochastic force on its growth rate, the population will, not surprisingly, be more variable than if there is no stochastic forcing. However, depending on the relative strength of the stochasticity, the population will form qualitatively distinct patterns. So, going back to the example of the fungal pathogen on the farm, suppose that we are dealing now with 50 distinct populations of the pathogen (i.e., 50 farms), each of which has precisely the same average growth and death rate. If there is a small stochastic force added to the growth rate of each of the populations and the populations persist for a long time (long enough to reach carrying capacity with no stochasticity added), some of the populations will be near their carrying capacity but others will be far from their carrying capacity. Thus we expect a distribution of population sizes something like that illustrated in Figure 5.4, where more than 50% of the populations reach the highest population size and smaller proportions of the populations reach lower values, but there are hardly any extinctions. With regard to the soil pathogen, this means that more than half of the farms will have high levels of the pathogen but a small proportion will experience low levels of the pathogen. However, all the farms will have at least some level of the pathogen. The interesting feature is that as the stochastic forcing becomes large, the distribution of population sizes will be bimodal, with many populations experiencing extinctions (or very low values), many close to the higher values, and fewer with intermediate values, as shown in Figure 5.4. What this means for the distribution of the soil pathogen in the 50 farms is that in many farms the pathogen will be at very low levels or disappear completely, in many other farms it will reach the highest levels, while in a smaller proportion of the farms, the pathogen population will reach intermediate levels (Figure 5.4). Thus, the general distribution of population sizes is qualitatively altered by a simple addition of a stochastic force.

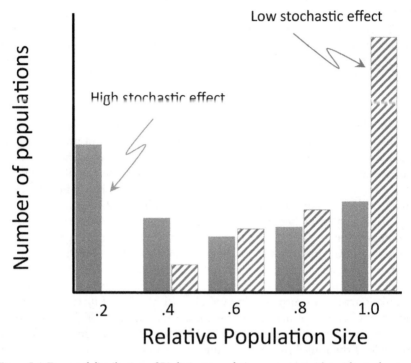

Figure 5.4 Expected distribution of 50 distinct populations growing with an identical average growth rate, but with two distinct levels of stochastic forcing

Taking a slightly more complicated example, the Allee effect is a well-established principle in ecology and evolutionary biology, wherein the size of a population is not simply a negative impact on the population growth rate, but at low densities can have a positive effect. The most evident, perhaps trivial, example is in sexual organisms, where a population of a single individual is, by definition of sex, impossible to maintain. Other less trivial examples are common in the ecological literature. For example, the introduction of insect parasitoids as biological control agents has been difficult partially because of this effect – the introduction of a relatively small number means that the probability that males and females get together to mate is reduced.[5] Schooling fish, herding mammals, flocking birds and many other examples are potential examples of the operation of this principle.

How does the addition of stochasticity to the basic Allee effect alter its basic pattern? One might expect that since there is some critical population density below which the population is likely to go extinct, adding a stochastic factor will cause populations to cross that critical point quite easily, and this is precisely what happens, as shown in Figure 5.5. With low stochastic forcing most of the populations reach relatively high population numbers while none reach low

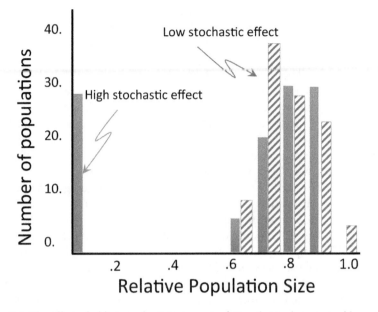

Figure 5.5 The effect of adding stochasticity to a population that is characterized by a strong Allee effect

values or extinctions. On the other hand, with high stochastic forcing a high proportion of the populations to go extinct (or reach very low values), another high proportion reaches high population levels and none reaches intermediate levels. Note that the qualitative result is basically the same as for a population that does not have an Allee effect, but the effect is much stronger.

In both of the preceding examples, adding a stochastic factor tends to generate an enhanced probability of local extinction. Recalling the basic ideas of metapopulations, we see here a clear mechanism for creating the local extinction pattern that is the hallmark of metapopulation dynamics. It is certainly not the only potential mechanism, but it is a very simple idea. A population that is completely deterministic, that is, without any stochastic factor, may very well exist in perpetuity wherever it is found. But if stochasticity is added, suddenly that same population may go extinct, probably only in places where it is relatively rare to start with, but given that many populations may exist as metapopulations to start with, such an effect could be significant.

It is clear that the ultimate fate of a population will be determined by both deterministic and stochastic forces. Using the very elementary considerations of population dynamics that we have been employing in this chapter, certain relationships emerge that may seem odd at first glance. Suppose, for example that we have a simple population growing according to two rules: at low densities it grows approximately exponentially and at high densities it feeds back on itself (e.g., may eat so much of its food source that it actually depresses its own population). We can imagine the dynamics of this population with a single measurement, the rate of reproduction of an average individual. If that rate (call it R) is too small, the population will not grow at all (obviously it needs to be greater than 1.0). However, if it is too large, the population will tend to grow so fast that it will consume too much of its own food and begin to decline, perhaps even go extinct. While the details are not all that important, the particular example we construct here sees the population as having an R somewhere between 1.0 and 4.0 (below 1.0 the population declines because of too few individuals produced, and above 4.0 it consumes so much food that its newborns have nothing to eat). Thus the population, from a strictly deterministic point of view, will exist in perpetuity as long as R is between 1.0 and 4.0.

We now ask what will happen if environmental noise (i.e., stochasticity) is brought into the system. Each year (or day or hour, depending on what sort of organism is under consideration), there is a certain probability that the population will become extinct. It is easy to calculate that probability with a simple simulation. In Figure 5.6a we show the basic qualitative results (different models will give different particular results, but the basic qualitative results is the same). Note that generally as the stochastic effect is increased the probability of extinction increases, not a surprising result. With added stochasticity, as we saw earlier (Figure 5.3), the population tends to fluctuate randomly, and, if the fluctuations are large enough, the population will descend below the zero mark, which means extinction. Also note that the transition from low to high probability of extinction as stochasticity increases happens quite abruptly and that the higher the R value is, the less stochasticity is needed to reach that transition (Figure 5.6a). However, it is also the case, that with a very low R value (e.g., R = 1.1, Figure 5.6a)

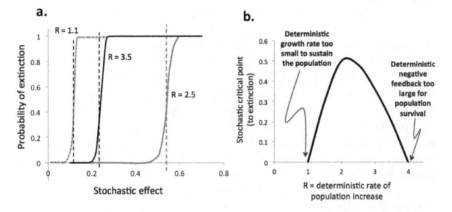

Figure 5.6 Results of adding stochastic forces to the simple model of density-dependent population growth. a. Extinction probability increases as the stochastic effect becomes larger, and the point at which it switches from low probability to a high probability is rather dramatic (indicated by the dashed lines), and depends on the value of the growth rate, R. b. The position of the stochastic critical point as a function of the deterministic rate of population increase (R). Note the interplay between deterministic and stochastic forces here.

the level of stochasticity needed to reach the transition point to a high extinction probability is also very low.

Figure 5.6b shows the relationship between the stochastic critical point where the extinction probability increases dramatically, and the value of R, the reproduction rate of the population. If the deterministic growth rate of the population is near its own critical point (either 1.0 at the low end, or 4.0 at the high end), only a small amount of stochasticity causes rapid extinction. However, at intermediate values of R, it takes a rather large amount of stochasticity to cause extinction. So we see the intermingling of both deterministic and stochastic factors determining the fate of the population.

Predator/prey systems and stochasticity

A particularly instructive example of the intermingling of deterministic and stochastic forces is the behavior of a predator/prey system with simple density dependence on the prey population. Recall from Chapter 2 that a graph of prey versus predator (the phase plane) appears as a spiral moving toward the unique equilibrium point, which implies that the inevitable oscillations of predator/prey systems over time will dampen to single values of predator and prey (Figures 2.20c and 2.20d). There is something of a conundrum here in that what appear to be the simplest biological assumptions (predator eats prey, and prey has an external controlling force, leading to its carrying capacity) lead us to the conclusions that permanent oscillations are not possible (all oscillations will dampen to a fixed point). As described in Chapter 2, the persistence of oscillations in both experimental and natural systems suggests that something must be lacking in the underlying theory (i.e., permanent, non-damped, oscillations seem to occur, at least sometimes), so we can add the idea of predator satiation (a functional response) and arrive at the situation in which oscillations are permanent. In the limit, predator and prey oscillate with respect to one another in a completely predictable pattern that repeats itself in a fixed period of time, a limit cycle.

Levins[6] long ago noted that simply adding a stochastic force to the basic predator/prey system (with density dependence for the prey) can give rise to a pattern that, while not formally a limit cycle, behaves as if it were. Consider the trajectory illustrated in Figure 5.7a, a damped oscillation in which predator and prey oscillate with respect to one another, but spiral in to a fixed point. If we now add a random factor such that any point on that spiral is perturbed by a fixed amount, what will happen? Without thinking too deeply, it might seem that the system will just become more "fuzzy," a distribution of points (predator/prey pairs of values) that are more or less around that original spiral. However, something rather more interesting happens. Take a particular point on the outer trajectory of the spiral, far from the equilibrium point, and draw a circle of some radius around it (Figure 5.7a). Now, presume that the original point will be perturbed in a random direction in the predator/prey space. You will notice that the number of points that tend to fall "outside" of the original trajectory is slightly larger than the number that will fall "inside" of the trajectory. This means that the stochastic effect will be to push the trajectory away from the ultimate equilibrium point. Now do the same thing but in the inner part of the spiral, close to the equilibrium point (Figure 5.7b). You will notice that the number of points that tend to fall "outside" of the original trajectory is much larger than the number that will fall "inside" of the trajectory. This means that a point located nearer to the equilibrium point, on the spiral trajectory, will be even more likely to be perturbed away from that equilibrium point than when it was located further away from the point (Figure 5.7c). The consequence is (1) a stochastic force will generally cause the system to move away from the

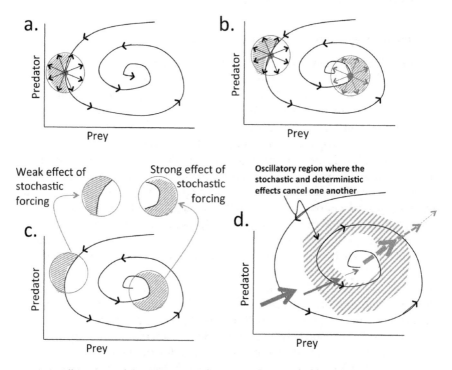

Figure 5.7 Illustration of the generation of stochastic limit cycle-like behavior in a predator/ prey system. a. A typical spiral form of the damped oscillations between predator and prey, where a single point is perturbed in a random direction. The small circle indicates the positions that the perturbations could take. It is evident that there are more points on the outside of the original spiral trajectory (indicated by the shaded part of the small circle). b. Repeating the random perturbation of the system for two points on the original spiral trajectory. Note that the circle nearer to the equilibrium point (circle on the right) has a much larger shaded region; that is, the region where the stochastic perturbations are away from the equilibrium point. c. Isolating the two exemplary perturbation points of (b) to see the relative sizes of the shaded areas. d. The ultimate consequences of a deterministic effect (solid arrows) that decreases the closer it gets to the equilibrium point and a stochastic effect (dashed arrows) that decreases the further away it gets from the equilibrium point. A kind of balance between the deterministic effect and the stochastic effect creates a region in which the system is likely to be found, with oscillations that appear to be like a limit cycle.

equilibrium point, and (2) the effect of that force will be stronger the closer the system is to the equilibrium point. Thus, there will be a balance between the deterministic force, the original force pushing the system toward the equilibrium point, which is stronger the further it is away from the point, and the stochastic force, the random perturbations that push individual points away from the equilibrium point, with a stronger force when near that point than when far away from it. The consequence is a tendency to move deterministically toward the equilibrium point and stochastically away from the point, generating a "donut-like" shape within which the system will tend to be found (Figure 5.7d). That donut area looks very much like a limit cycle, but, of course, formally, it is not. This is a simple way to generate what are, effectively, permanent oscillations of a predator/prey system.

The message here is clear, and it is not only about predator/prey cycles. The qualitative understanding of an ecosystem may be different depending on whether stochasticity is taken into account or not. As we noted in Chapter 4, sometimes it is just assumed that chaotic systems are effectively stochastic and the question becomes either of how to characterize the stochasticity itself, or how to minimize its effect in an empirical data set. However, a dedicated group of theoretical ecologists remains committed to understanding the interrelationship between chaos and stochasticity. To give a sense of the sorts of questions they ask, it is useful at this point to repeat the exercise we introduced in Chapter 4, where we query a farmer about his or her pest problems.

The interrelationship between chaos and stochasticity

To construct this example in a didactic fashion, let us suppose that the farmer's experience is relatively short (about a decade), and what she has experienced on her farm is an oscillatory situation in which one year the pest is abundant, the next year not so much, the next year abundant again and so forth. Furthermore, her experience suggests that big oscillations alternate with small ones. After further questioning, the pest control technician concludes that the oscillations go from 8 individual pests per plant, to 6 per plant, to 9 per plant, to 3 per plant, which is to say the big oscillations go from 9 to 3 and the small oscillations go from 8 to 6. She seems convinced that this pattern is unique and will continue forever. Note that she is working in discrete time and with discrete numbers – a very important fact. Her informal projections, then, will be a series like this: 6, 9, 3, 8, 6, 9, 3, 8 and on and on. Can we use standard modeling techniques to reproduce what she expects, or are her expectations not in accord with any known ecological theory? The technician comes up with a simple model – and it is important to acknowledge that this model system is in discrete time (that is, from year to year) – but using a continuous variable (that is, we could have 8.6 individuals or 3.1 individuals, not exactly realistic, but then again, models always abstract nature). Fitting the model to the available data, she concludes that the model represents an excellent fit to the data provided by the farmer and thus is confident that it will predict the

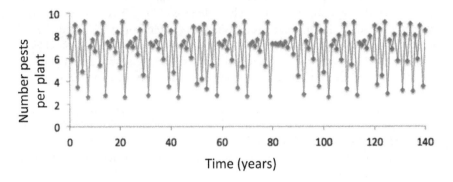

Figure 5.8 Results of a population growth model with a continuous variable (for number of pests per plant). Note that the model fits the empirical data of the farmer's experience for the first 8 years but then it deviates dramatically.

future adequately. But if we run her model for more than 100 years, we get the result pictured in Figure 5.8.

Clearly, the model predicts something different from the farmer's experience. Note that the model itself was adjusted initially to the farmer's experience, and it fits the initial data very well. So, can we trust the predictions, which says that only rarely will we see the general 4–year cycle thought to be true by the farmer? Who is right: the model projections or the intuition of the farmer?

But perhaps the farmer notes that our model framework is continuous (i.e., admitting fractions of individuals) while in nature she observes integers. In other words, she never observes 5.3 individuals. What if we just rounded the projections of the model? At least that would correspond to reality more closely since individual insect pests are in fact integers. If we do that (take the results of the model and round them to the nearest integer), we obtain the results illustrated in Figure 5.9.

Thus, we see that the continuous model, when just rounded to the nearest integer, basically produces a similar result to the non–rounded continuous model itself. That is, the system seems to be a chaotic system. So, are we now confident in telling the farmer that, indeed, this is a chaotic system and she cannot really predict with any level of confidence what the next year will hold, based on her experience over the last few years?

But perhaps the farmer then notes that she really didn't mean to say that the number of pests every year were exactly 8, 6, 9 and 3, but rather that these were the average values she observed per plant, and indeed different plants had different values each year, but, on average, each year the values were either 8 or 6 or 9 or 3. Effectively what she is saying is that our model is wrong not because of how we fit it to the data initially, but because we presumed there was no stochasticity (no inherent variability) in the system. Why not, she says, project a "distribution" of values, and take the mean of that distribution, and round that to the nearest integer, as she did? If we do that, we obtain the results in Figure 5.10, a confirmation of the farmer's intuition!

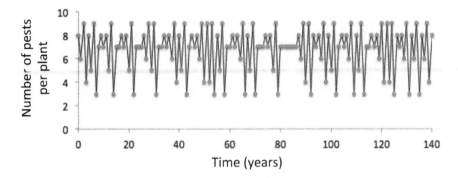

Figure 5.9 Results of a population growth model rounding the continuous variable (pest population size) to its nearest integer

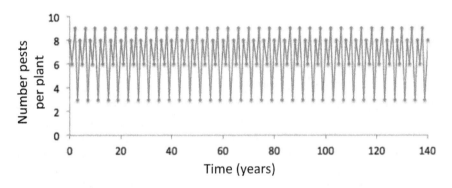

Figure 5.10 Results of a population growth model adding stochasticity and then rounding the continuous variable (pest population size) to its nearest integer

So we see that the combination of a model that is inherently chaotic, plus the addition of a reasonable stochastic framework, actually predicts what the farmer had stated from the beginning. And we see the absolute need of the continuous, integer and stochastic elements of the model system to represent this (imagined) reality.

We must add here, that our example was contrived to make a specific point. It did not have to turn out that way. For example, something different happens if we begin with the initial observations that for 5 years we have the average number of individual pests per plant as 5.00, 9.58, 1.56, 5.04, 9.57, 1.56, which, after rounding the numbers, gives us 5, 10, 2, 5, 10, 2. Thus there seems to be a repeating 3-year cycle. And it is easy to find a basic model that makes that prediction, before adding any stochasticity. That is, the farmer's initial suggestion that there is a three-stage oscillation is reinforced by a fitted model that predicts a continuous number (e.g., 9.58 or 1.56), which can be rounded to correspond to the reality of animals being integers, and it produces precisely what the farmer says she expects. However, in this case, with this simple model, if we first add

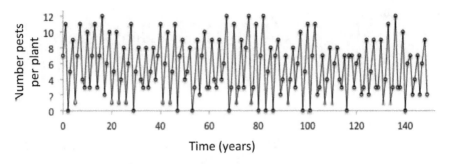

Figure 5.11 Results of a population growth model adding a stochastic force and rounding to the closest integer, when the model is initially adjusted to predict precisely a 3-year cycle

the stochastic element and then round to the nearest integer, we can obtain a trajectory that looks very much like chaos, as shown in Figure 5.11.

The chaos/stochasticity tapestry

What we see from the previous simple examples is that there can be an intimate relationship between chaos and stochasticity. Part of the reason this is so, at least in the examples of the preceding section, is that there are frequently places in the abstract space of a model where the output of the model is highly dependent on the exact values of its parameters. A particular form of behavior, say chaos, characterizes the model, but a very small change in one of the model parameters (the rate of population increase, for instance) may push it into a zone of very regular behavior (e.g., Figure 5.10). This is a common effect that can be reproduced in many theoretical frameworks, but we must ask to what extent it might actually apply in the real world, and here we have a body of experimental work that has been analyzed in detail, that suggests that the real world may, indeed, be a "tapestry of chaotic elements woven together by stochasticity."

The relevant "real world" example comes from a small beetle that is a pest of stored grain, *Tribolium castaneum* (Figure 5.12a). This small insect is easy to cultivate in small population containers (Figure 5.12b), and therefore has been used as a model organism for some of the most interesting studies in population dynamics.[7] A relatively simple mathematical model does extremely well at predicting the population sizes of the beetles in these small containers. For example, one form of the model predicts that the population should oscillate between two densities, and that between those two densities it should experience an unstable point (the whole relationship should look something like a saddle point (refer to Chapter 2, Figure 2.25, if necessary, to understand the basic behavior, although details of the present model are quite distinct from the example in Chapter 2). So the expected population behavior would look something like that pictured in Figure 5.13.

Figure 5.12 Tribolium castaneum and an example of its environment. a. Adult beetle (source: Wikipedia Commons). b. Incubator containing vials in which the *Tribolium* populations were held in the laboratory of Robert Constantino (courtesy of Dr. Constantino).

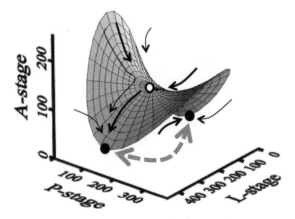

Figure 5.13 Diagrammatic illustration of the expected behavior of a population of beetles in three dimensions (A = adult densities; P = pupal densities; L = larval densities). The imposed solid is only for explanatory purposes (no such structure actually exists in the model). Note the existence of an unstable saddle point (the white dot in the center of the surface) and the two points in a cycle (the black dots), with the bold dashed double-headed arrow indicating switching back and forth between the two equilibria.

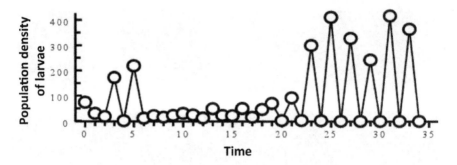

Figure 5.14 Larval population density of *Tribolium castaneum* over time, illustrating the two major modes of population behavior, early on approaching the unstable point and lingering there for a long period of time, then moving into the expected two-level oscillations (not exactly the same two points, but one high, followed by one low, followed by one high, and so forth)[8]

Source: Data from Jillson, 1980, adapted from Cushing, 2007, with permission, Springer.

Using real population data based on a laboratory experiment done in 1980, the population numbers over time seemed at first odd, as shown in Figure 5.14. However, with the aid of the model that predicted a two-level attractor (with populations jumping from high values to low values every 2-week period), those data make perfect sense (for a qualitative interpretation, see Figure 5.15). The population approaches the unstable point and lingers there for some time, but eventually moves toward the two-point oscillatory framework, conforming nicely to the expectations of the theory.

However, in another experiment, the population seemed to correspond to a chaotic attractor, as illustrated with the raw data presented in Figure 5.16a. Much as we argued earlier, that attractor has some regularity associated with it, in particular, various repeating cycles. The unusual aspect of this case is that the underlying theory, while clearly predicting a chaotic attractor for the situation, could not generate a particular feature of the empirical data. Looking carefully at the data, it became obvious that there was a repeating cycle of length six. Yet, the model, based on a non-stochastic approach, could not replicate this feature of the chaotic system (the six-point repeating cycle – illustrated in red in Figure 5.16b). Rounding the continuous model to its closest integer did not help – no six-point cycle could be found (with that particular parameter setting).

Then the researchers decided to apply a stochastic force to the model and, *voilà*, the new model, also a chaotic attractor, was able to produce the six-point cycle that the real data had suggested (Figure 5.16c).

In a related analysis, this same research group carefully analyzed the chaotic attractor of this system and identified four basic modalities that could be recognized within the attractor (a six-point cycle, an eleven-point cycle, a three-point cycle and an eight-point cycle). They conclude that it is the "transient but recurrent cyclic patterns generated by [chaos], woven together by stochasticity,

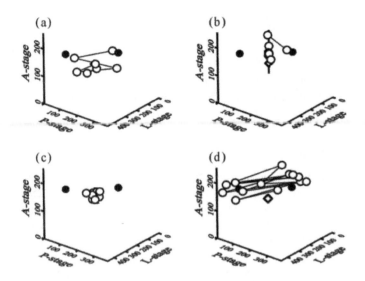

Figure 5.15 Three dimensional plots illustrating the time series for A = adults, P = pupae, and L = larvae of *Tribolium castaneum* in the Jillson experiment (as shown in Figure 5.13). The black dots indicate the two points of a two point cycle, and the white diamond indicates the unstable point. a. The initial behavior of the system far from either the unstable point or the two-point attractor (see Figure 5.12 for a qualitative interpretation of the dynamics and Figure 5.13 for the actual data of the larvae). b. The approach of the system toward the unstable equilibrium point. c. The population "lingering" near the unstable point for 40 weeks (twenty 2-week time intervals). d. The population finally moving to the expected two-point cycle, where it jumps back and forth between the two values.

Source: Adapted from Cushing, 2006, with permission, Springer.

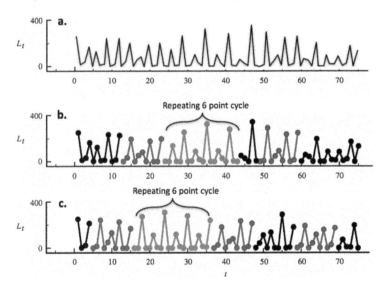

Figure 5.16 Results of adding a stochastic factor to a basic population model.[9] a. The empirical data illustrating a chaotic attractor with some basic structure. b. The empirical data shaded to show the repeating six-point cycle (also visible in lighter shade is an 11-point cycle). Note that the cycles are not perfect cycles, but rather generalized modes. c. The model predictions after adding the stochastic force in the model.

Source: Data extracted from Henson et al., 2001.

that distinguishes chaos as it is manifested in noisy discrete-state population systems."[10] This analysis reinforces our discussion of chaotic systems (in Chapter 4) and our main point, that there is indeed pattern in chaos. However, as illustrated with the study of chaos in the *Tribolium* population, it may take some effort, perhaps with some sophisticated analysis, to recognize that pattern.

Final comments

In this chapter we emphasize the effect of stochasticity as a force unto itself, not just the annoying variability that we seek to minimize in experiments, nor the basis of a statistical test. We see how simple population dynamics can be qualitatively altered by the addition of stochasticity and how the outcome of simple models of population interactions can change from the expected damped oscillations in a predator/prey model to permanent oscillations. Finally we introduced the especially important topic of the relationship between chaos and stochasticity. This last point is especially important since one of the most common observations of chaotic systems is that they are difficult to distinguish from randomness. If chaos is difficult to differentiate from randomness, is it not as interesting and important to ask whether randomness is different from chaos? The essential tapestry that emerges from the interweaving of chaos and stochasticity remains a major research problem in complexity theory.

Notes

1 Vandermeer et al., 2010.
2 There is a rich literature about this debate. See, for example, Strong, 1986, for an excellent summary.
3 Lewontin and Cohen, 1969.
4 Levins, 1969.
5 Hopper and Roush, 1993.
6 A point he made verbally to JV several times, probably first in 1969.
7 Thomas Park originally conducted a series of detailed competition experiments with these organisms (Park 1954: Neyman et al., 1956), and more recently the "Tribolium" team applied modern mathematical and statistical techniques to the system (Costantino et al., 1997, 2005; Henson et al., 2001).
8 Data from Jillson, 1980, adapted from Cushing, 2007.
9 Henson et al., 2001.
10 King et al., 2004.

6 Coupled oscillators

An odd kind of sympathy

In the mid-seventeenth century, the world's second most famous physicist reported on a strange behavior of pendulum clocks. Having taken on the challenge of constructing clocks of sufficient accuracy to be used in determining longitude on long ocean voyages, Christian Huygens invented the pendulum clock, which he thought would be useful for this challenging seventeenth-century problem. Lying sick in bed in 1665, he noticed that two of his pendulum clocks, mounted on the wall, were oscillating approximately 180 degrees out of phase with one another. Curiosity drove him to change the pendula to be swinging in phase, and within a half hour, once again they had taken up an antiphase coherence with one another. Repeating this experiment, probably many times, he convinced himself that this phenomenon was a real element of pendula. He wrote to a confidant that his two clocks displayed "an odd kind of sympathy . . . [when] suspended by the side of each other."[1] A reminder of this episode can be encountered on Leidsestraat in Amsterdam, where a mosaic of him with his tell-tale clocks can be seen (Figure 6.1).

What Huygens discovered has become the foundation of a variety of applications in many sciences and a variety of engineering applications – when coupled, sometimes very subtly, two oscillators will come to "inform" one another about their oscillations and attune to each other. Many illustrations of this phenomenon can be seen in abundant video clips, and applications from signaling fireflies to electronic oscillators are well-known.[2] For ecology, as we noted in Chapter 2, one of the most evident of ecology's rules is the consumption of one organism by another, whether bacteria in the soil or parasitoids attacking pests. This trophic connection is, at its most foundational level, oscillatory. It is in this sense that all ecological systems are collections of coupled oscillators, and it is not surprising that this metaphor has been applied to coupled ecological systems generally, and especially to classical trophic systems (predator/prey, herbivore/plant, etc.).

Figure 6.1 Mosaic on front of building at Leidsestraat 88, Amsterdam

Consumer/resource oscillators and weak coupling: a basic pattern

The simple observations of Huygens in the seventeenth century serve to enlighten ecologists of the twenty-first. Trophic structures, the consumption of organisms by other organisms, are, by elementary observation (see Chapter 2) oscillators – large concentrations of resources lead to large numbers of consumers, large numbers of consumers lead to lower concentrations of resources and so on. Consider, for example, the various elementary oscillators in Figure 6.2a (the creative reader probably can think of others). These oscillators can be coupled, that is, they may be connected to one another as were the pendulum clocks of Huygens. One of the ways of coupling them is by allowing them to eat the "wrong" food – Consumer 1 preferring Resource 1 but sometimes eating Resource 2 and Consumer 2 preferring Resource 2 but sometimes eating Resource 1. This is the situation diagrammed in Figure 6.2b, with the two consumers sharing two resources. Here we see that the coupling of the oscillators is

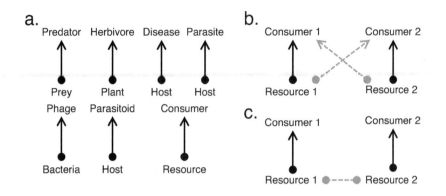

Figure 6.2 The variety of trophic oscillators common in nature. a. Alternative forms of the basic trophic "atoms" of the ecosystem. b. The elementary form of resource competition, applicable to any of the alternative forms, here illustrated with the generalized "consumer" and "resource." c. An alternative coupling of the basic consumer resource oscillator, through competition between the two resources.

through the joint utilization of the two resources, sometimes summarized as "the mechanism of competition is joint resource utilization," a very common situation in both theoretical thinking about ecology and in nature. However, there is another way of coupling these oscillators that makes biological sense. It could be that the resources themselves are in competition for some other resource, but that resource is not stipulated. Rather, the "phenomenon" of competition is simply stated as the impact of one resource causing a reduction in the growth rate of another resource (note that we are thinking here mainly of resources that themselves are self-replicating – not really a necessary assumption, but a convenient one). This situation is illustrated in Figure 6.2c.

Thus we may conceive of competition operating at either a "mechanistic" level, when the resources being competed for are stipulated, or a "phenomenological" level when they are not. In either the mechanistic or the phenomenological approach we have two consumer-resource oscillators that are coupled (either through consumption, as in Figure 6.2b, or through competition between the resources, as in Figure 6.2c). The consequences of either of these sorts of competition are interesting, as shown in Figure 6.3. When coupling is through the joint utilization of resources by the consumers (mechanistic competition coupling), oscillations that begin out of phase gradually come into phase with one another (Figure 6.3a). Contrarily, when coupling is through the phenomenon of competition between the resources (phenomenological competition coupling), oscillations that begin in phase, gradually come to an anti-phase coordination with one another (Figure 6.3b).

Inspired by Christian Huygens's original observations, in Figure 6.4 we present a physical metaphor that may be useful. We see two pairs of pendula. Each

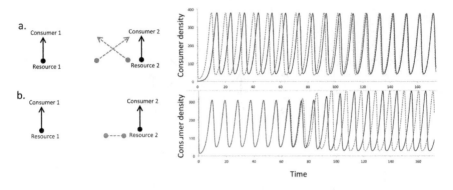

Figure 6.3 Simulations of weakly coupled predator/prey oscillators, with the y-axis plotting the population density of consumer pairs (bold trajectory is consumer 1, light trajectory is consumer 2). a. The basic competitive connections of mechanistic competition. b. Phenomenological competition. In (a) populations are initiated in an "anti-phase" coordination and gradually come into phase with one another. In (b) populations are initiated in an "in-phase" coordination and gradually come into "anti-phase" coordination with one another.[3]

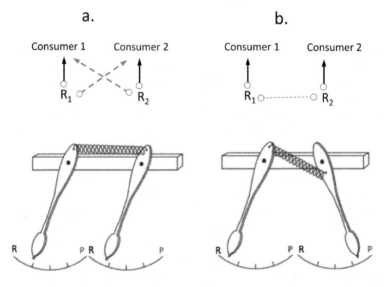

Figure 6.4 The elementary coupling of consumer-resource systems and the qualitative outcomes. (a-top and b-top) Two consumers eat two resources, with arrowheads indicating a positive effect and small circles a negative effect. The form of coupling is (a-top) that the consumers eat each other's resources (and thus become competitors with one another), or (b-top) that the resource items compete with one another. (a-bottom) Physical metaphor of coupled pendulums, with a metaphorical spring connecting them in a way that causes them to become phase synchronized, which is what happens with weak coupling as in panel a-top. (b-bottom) Physical metaphor of coupled pendulums, with a metaphorical spring connecting them in a way that causes them to become anti-phase synchronized, which is what happens with weak coupling as in panel b-bottom. The scale at the bottom illustrates how the metaphorical pendulums oscillate between numerical dominance of consumer and numerical dominance of resource. The two forms of coupling generally result in either in-phase (bottom left) or anti-phase (bottom right) coordination of the oscillators.

Source: Figure modified from Vandermeer, 2006.

pair has a spring coupling them. Mechanistic competitive coupling is similar to attaching a spring connection (a weak spring) to two pendula in the form as illustrated in Figure 6.4a, while phenomenological competitive coupling is similar to attaching a spring connection in the form as illustrated in Figure 6.4b. The spring coupling as in Figure 6.4a leads to the ultimate result that the two pendula come to oscillate in the same phase. The spring coupling as in Figure 6.4b leads to the ultimate result that the two pendula come to oscillate in an anti-phase cohesion. Staring at Figure 6.4 for a few minutes will convince the reader that these two forms of coupling indeed will lead to such phase correlations. And, it is clearly the case that when coupling is like Figure 6.4a the oscillations will usually come into phase with one another, while when the coupling is like Figure 6.4b the oscillations will usually come into an antiphase correlation with one another.

These very regular results emerge from coupling consumer resource systems in a particular way (Figures 6.3 and 6.4), but only when the coupling is weak. As we increase the strength of the coupling, something else happens. As is well-known in coupled oscillators in many fields, various forms of chaos emerge as the coupling increases in strength. The details are probably less important for the current book than would merit spelling them out. Suffice it to say that the nature of chaos can be quite complicated when oscillators are coupled strongly, yet the underlying cohesion (whether they are in phase or out of phase) remains a feature that can sometimes be seen as part of the "structure of chaos." Much like the example of Chapter 5 where different cycles could be seen within a chaotic attractor, similar cycles can sometimes be seen when chaos emerges from coupled oscillators.[4] An excellent example is seen in the so-called teacup attractor.

Oscillatory structure in chaos: the teacup

One of the most ubiquitous situations in nature where oscillators are coupled is in the structure of a simple trophic chain, as illustrated in Figure 6.5; lizards eat beetles and beetles eat aphids. Such a simple and ubiquitous arrangement can result in some remarkable complexities. Coupling these two oscillators leads,

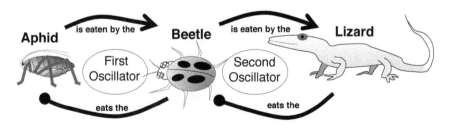

Figure 6.5 Cartoon representation of a three trophic system (trophic level 1 = aphid herbivore; trophic level 2 = lady beetle predator of aphid; trophic level 3 = lizard
predator of lady beetle), as two coupled oscillators

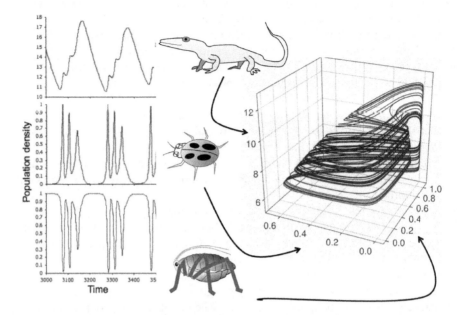

Figure 6.6 The tri-trophic model of Hastings and Powell, a chaotic attractor. On the left, each of the variables over time, illustrating the clear signal of two separate attractors; on the right, all three plotted in three-dimensional space, illustrating the chaotic nature of the system.

with proper parameter values, to a very specific chaotic attractor, known generally as the "teacup" attractor due to its shape in three dimensions. In Figure 6.6 we illustrate the time series for each of the three elements plus the three-dimensional portrait of the attractor. The behavior is well-known to be chaotic, but it is important to note, as we emphasized in Chapter 3, there is a clear structure to the system. But more importantly, here we can see the imprint of where the attractor comes from in the first place. There are basically two types of oscillations that, in a sense, share the space: long-term oscillations between top predator (lizard) and predator (beetle), and short-term oscillations between predator (beetle) and herbivore (aphid). So even though the overall system is chaotic, its underlying structure of two coupled oscillators is nevertheless evident, and the relative amplitude and frequency of each of the oscillators is clear within the chaos.

Confronting Gause's principle with oscillations

A key central concept in ecology is that no two species can occupy the same niche (see Chapter 2). However, this idea becomes dramatically altered in the context of oscillations. Consider, for example, the situation as presented in Figure 6.7. It is a common, perhaps universal pattern. The herbivore (here an aphid) has both

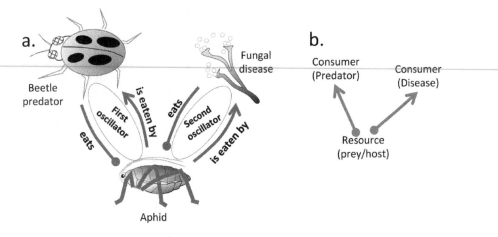

Figure 6.7 Common, almost universal, arrangement of an organism simultaneously consumed by a predator and a disease. Since the two consumers (the lady beetle predator and the fungal disease) in (a) consume the same thing (the aphids), they live in the "same niche" (sort of) and thus, according to basic ecological theory, cannot coexist. (b) The basic arrangement in a more abstract presentation.

a predator (the beetle) and a disease (the fungus). In fact it is the case that all living organisms have diseases (we know of no exception) and, with the exception of elephants and large whales, all living organisms face predators in their native environment. Yet the elementary principle that "no two species can occupy the same niche" suggests that such an arrangement, in and of itself, is impossible. So, the fundamental principle of ecological competition says that no two species can occupy the same niche, seems to face the problem that it is almost universal that organisms live simultaneously with diseases and predators, which means that two species are indeed using the same resource.

This apparent contradiction is elegantly resolved simply by acknowledging that the arrangement is one of coupled oscillators. Such an acknowledgment means that if the underlying idea generating this theory is only slightly relaxed, the conclusion itself disappears.[5] We revisit this issue in Chapter 8, but for now we note that if the populations are allowed to oscillate, what one might expect qualitatively in fact happens: when the disease is at a low intensity, the predator can invade the system, but when the predator is at low density, the disease can gain a foothold. This is a bit of an oversimplification, but it is essentially the idea. In agroecological applications it is frequently the case that a pest manager seems to be confronted with the problem of choosing a predator or disease as a biological control strategy. Yet, if we take our cues from nature, predator and disease are problems that all organisms must face, simultaneously.

Let's then imagine that we have a situation of a pest (an aphid) that is attacked both by a predator and a disease (Figure 6.7). Looking at this situation theoretically, we have two oscillators (predator/prey and disease/host, the latter of which is the same as the prey of the predator), connected together in the basic form of two species competing for the same resource (Figure 6.7b). If we develop a simple disease/host model using the basic form normally used for analyzing disease systems, and a simple predator/prey model using the basic form normally used for analyzing predator/prey systems, and set them up separately to generate unstable oscillations, the outcome for both systems separately will be the eventual extinction of the disease or the predator and, in both cases, the increase to its carrying capacity of the prey (or host), in this case the aphid (or pest). In other words, the pest goes uncontrolled. However, a remarkable thing happens when we connect the two subsystems. With the same parameters in the model, but linking the two subsystems so that they both have the same the prey or host (the aphid), the outcome is a system that is forever oscillating (and is a chaotic attractor), but where the prey or host (the aphid) is kept under control.

However, the story here can be bit more complicated. For example, we can ask, what is the consequence of the predator eating diseased individuals? If we presume that the predator does not get its "full" energy package when it eats an infected individual rather than an uninfected one, the outcome of the interactions can be quite different. While there is a basic conclusion that two control agents tend to control a pest better than either agent alone, the structure of the control can be quite distinct when the penalty for eating diseased individuals is significant.[6] In particular, if we presume that the penalty for eating a diseased prey is about 90%, two alternate states, both chaotic, are generated from the model. That means that depending on where we start running the model, we will end up in one or the other of these two alternate states (the yellow and red trajectories in Figure 6.8a). In one of these states, the predator seems to dominate the oscillations with the prey (Figure 6.8c) and in the alternative state, the disease seems to dominate the oscillations with the prey (or host in that case) (Figure 6.8d). In the dynamic modeling lingo, this situation represents a system with two basins of attraction separated by a separatrix (the border between the two basins). In our example, when the system starts close to the light shaded basin of attraction it eventually ends up in the light shaded chaotic attractor, but when it gets started closer to the dark shaded basin, it ends up in the dark shaded chaotic attractor. Interestingly, when we increase the penalty rate to 95%, the outcome is a single chaotic attractor that combines the two trajectories (or chaotic attractors) previously described (Figure 6.8b and 6.8e). What happens here is that increasing the penalty rate to 95% causes the oscillations within the attractors to expand to a point where they reach the separatrix of the other attractor, and the result is that the system oscillates in one basin of attractions for a while and then (when it reaches the separatrix) it moves to the other basin of attraction, and these continue back and forth forever, generating a pattern that combines the two patterns generated by the two alternate states when the penalty was only 90% (Figure 6.8b and 6.8e). This is yet another example where chaotic attractors can be seen to reflect something of their underlying structure, here the semi-independent contributions of predator and disease to pest control.

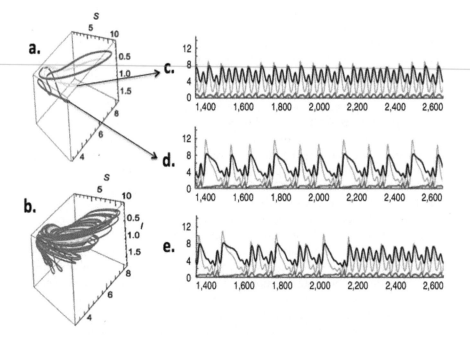

Figure 6.8 Results of simulations of the simultaneous action of a predator and a disease on a pest (S = susceptible pest, I = infected pest, P = predator). a. Results of the simulation when the penalty for eating infected individuals is 90%: alternative chaotic attractors, one darker and one lighter shade. b. Results of the simulation when the penalty for eating infected individuals is 95%, a single chaotic attractor where the dark and light shaded modalities can still be recognized. c. Time series of the chaotic attractor where the disease is the dominant control element. d. Time series of the alternative chaotic attractor where the predator is the dominant control element. e. Time series of the combined chaotic attractors where both modalities (predator control versus disease control) can be recognized. In the time series (c, d and e) dark grey shaded = susceptible pest; light grey shaded = infected pests (or disease); black = predator.

Source: Ong, T. W. Y., & Vandermeer, J. H. (2015). Coupling unstable agents in biological control. *Nature Communications*, 6.

Limiting similarity and species packing with oscillators

More in line with Huygens's original insights, some classic ecological questions gain significant enlightenment when cast in the form of coupled oscillators. An example is the famous Gausean principle that no two species can occupy the same niche, or formulated in more modern terms, two species with sufficiently large niche overlap cannot coexist in perpetuity. Long a persistent framework in ecology, it is really a problem of coupled oscillators (Figures 6.2 and 6.3), although not always recognized as such. The principle refers to two species that are either in the same or similar environments, or, more commonly, two species that share some resources (and thus compete for them). Elaborating with some insightful natural history generalizations, MacArthur and Levins (1967) noted that if the niches of two species overlapped too much, one or the other species would be eliminated, as the classical theory says. However, they also noted that selection would operate most strongly on those individuals located in the zone of overlap, favoring those not located in that zone. The result is obvious. If there is heritable variation in niche use (where "use" is really just a position on an environmental gradient), those individuals located in the zone where there is the most overlap (Figure 6.10) will tend to be eliminated. The consequence would be a reduction in the breadth of the niches of each of the species (Figure 6.10b), along with a reduction in the zone of overlap. Obviously, if there is enough reduction in the zone of overlap, there will be room in the environment for an additional species to move in (Figure 6.10c).

A bit of a footnote is worthwhile here. Although this theory of "limiting similarity" and "species packing" was originally formulated as an evolutionary argument, its utility as a conceptual framing is far broader. This is especially true when considering so-called novel ecosystems (i.e., ecosystems that contain components that have only recently come together, the most obvious example of which is the agroecosystem). Suppose, for example, that three invasive species arrive at the shores of an established ecosystem. When the first two species arrive, they displace two or three other species that had been there before. The question that emerges naturally is whether those two species form the sort of niche occupancy as pictured in Figure 6.9a or, alternatively, as pictured in Figure 6.9b. The consequences for the arrival of the third species are obvious. So, if the novel ecosystem (species 1 and 2 together) is fully "packed" as in Figure 6.9a, the next new arriving species will not be able to fit in. But if the novel ecosystem is composed of fully niche-separated species, as in Figure 6.9b, the next new arriving species will fit in fine, as it does in Figure 6.9c. Species packing then is not only an evolutionary phenomenon but also has ecological relevance.

Figure 6.9a, 6.9b and 6.9c illustrate a classical formulation of the "limiting similarity/species packing" relationship. It says, in very broad and general terms, if the niches of two species are sufficiently similar (as in, for example, Figure 6.9a), a third species cannot invade the community. There is a limit to the number of species that can fit in, based on the overlap in their niches. But if there is only a small amount of overlap (as in Figure 6.9b), it is possible to "pack" a third species into the system (as happened to produce Figure 6.9c).

There is a more dynamic way of envisioning this old framework. If we conceptualize the resource gradient (the abscissa of the graphs on the right-hand side of Figure 6.9a, 6.9b and 6.9c) in two broad zones, such that we have effectively two distinct resources, R_1, and R_2, located more or less at particular points, or regions, illustrated by the transparent vertical bars crossing through Figure 6.9a, 6.9b, and 6.9c, we can formulate the problem as a trophic problem. That is, for each of the resources we can imagine a consumer, or consumers, attacking them. Thus, we can represent the three cases of Figure 6.9a, 6.9b and 6.9c, with graphs of consumers eating resources, as shown in Figure 6.9d, 6.9e and 6.9f (note that

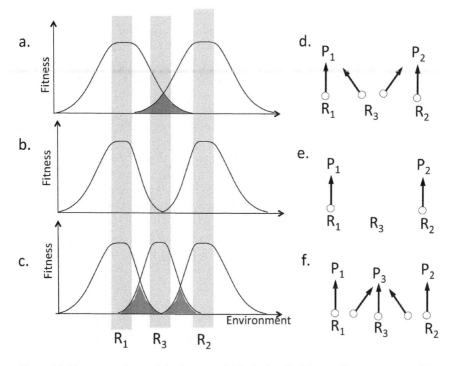

Figure 6.9 Elementary form of the Levins and MacArthur limiting similarity argument. The x-axis represents an arbitrary resource gradient, and the y-axis represents some measure of fitness for each consumer species. The intersections (shaded grey areas) represent the areas of intense competition between the two species. a. Selection favors the individuals that are not located in the intensely competitive zones where the niches overlap. b. The result is a reduction in the breadth of the niches of the two consumer species, effectively two separate consumer/resource oscillators. c. As the overlap between the two niches shrinks, enough niche space becomes available to allow a third consumer species to invade the system. Figures d, e and f are representations in the form of oscillators of the corresponding niche graphs of a, b and c.

the various panels in Figure 6.9 are paired, a and d representing the same situation, likewise the pairs b and e, and c and f).

Reformulating the classic idea in this way, we see that it is fundamentally a problem of oscillators, in particular, coupled oscillators. Framing the idea of limiting similarity in this fashion, we can envision the introduction of a third species into the system as if it were the agent of coupling the two oscillators. Adding this extra dimension complicates things considerably. Yet, within the framework of phase cohesion (when oscillators are either "in phase" or "out of phase," and if out of phase, how much out of phase), examination of the structure of ecosystems takes on a new and perhaps more enlightened form than earlier approaches that simply searched for equilibrium. For example, rather than asking simply whether a third species can invade, we might query the pattern of cohesion formed by the two initial species and how that pattern is transformed by the introduction of the third species. In other words, the classical ecological problem of limiting similarity and species packing can be envisioned as a problem of coupled oscillators. Viewing the subject in that fashion actually provides an important insight, namely that the third species can be added in two fundamentally different ways, as illustrated in Figure 6.10. First, the third species can be introduced as a resource that is competing with the other two resources although the mechanism of competition is not stipulated (Figure 6.10a; recall the situation in Figure 6.4b). Second, the third species can be introduced as a consumer that shares resources with the other two consumer species (Figure 6.10b; recall the situation in Figure 6.4a). We refer to the first case as "phenomenological coupling" because the third resource is added as a competitor, where the competition is introduced as a phenomenon in and of itself (we do not stipulate what it is that the species are competing for). On the other hand, we refer to the second case as "mechanistic coupling" because the third species is introduced as a consumer on the two resources, thus stipulating the mechanism underlying the competition among the three consumers.

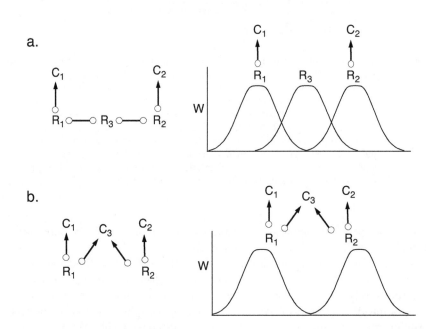

Figure 6.10 The two ways in which a third species can enter the ecosystem of two independent oscillators such that they are coupled through the third species. a. The invading species (R_3) is a competitor with the other two resource species. b. The invading species (C_3) is a predator on the resources of the other two species, and thus indirectly a competitor with them.

Consider first the mechanistic case (Figure 6.10b). Phase coordination will normally be in the direction of "in-phase" coordination with the two parent systems thus producing an especially strong competitive effect, since both C_1 and C_2 reach their peak values simultaneously with C_3 as a result of the inevitable phase coordination between C_1 and C_2 with C_3. If we put the uncoupled systems (before we add C_3) set in anti-phase coordination (we can do this arbitrarily since they are not coupled in any way before C_3 is added), and then add C_3, the effect is first to push the two parent systems into in-phase coordination. With the parent systems coordinated, even if each of the original consumers (C_1 and C_2) are only weak competitors with C_3, they exert a combined double effect on C_3 (because they are themselves in phase), such that C_3 will begin to decline. So, as we see in Figure 6.11, beginning with the parent system in anti-phase

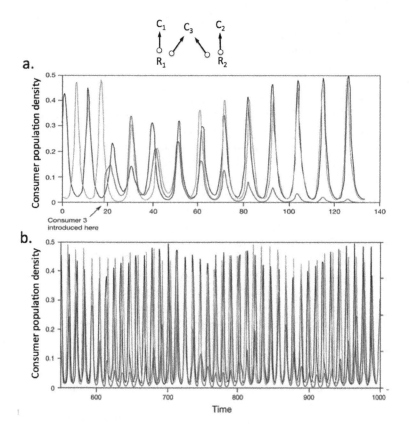

Figure 6.11 Control of an invasive species through an alternation of in-phase and anti-phase coordination of the two consumers. a. Time series shows basic process of consumer 3 forcing the other two consumers into synchrony. b. Long-term effects of invasive (intermediate shade) and permanent (dark and light shades) resident, where the cycle of anti-phase coordination of consumers 1 and 2 repeatedly emerges, to be repeatedly forced into in-phase cohesion. An obvious pattern with an obvious underlying biological mechanism is apparent (see Vandermeer, 2006, for details).

coordination, when C_3 is introduced it begins driving both C_1 and C_2 toward zero, but because of the coupling effect that leads to coordination of the oscillations, C_1 and C_2 rapidly come into phase with one another, thus driving C_3 toward extinction (the pattern is obvious in Figure 6.11).

We might think of this system as representing ecosystem resistance to an invasive species (with the two uncoupled oscillators as the base ecosystem and C_3 as the invading species). The phenomenon of synchrony, then, explains one way in which a community can resist an invasive species (i.e., first the invasive acts to synchronize the oscillations of the previously independent oscillators, and subsequently the coordinated behavior of those oscillators drives the invader to extinction). Note that the system itself is actually a chaotic attractor, emphasizing the fact that there is inherent structure in chaos (as repeatedly emphasized in Chapter 4).

If we modify this system just a little, such that there is a weak competitive effect between R_1 and R_2, the original system (the two parent systems, C_1/R_1 and C_2/R_2) will tend toward anti-phase coordination. We thus have a situation in which the invader will be excluded due to the synchronizing effect of the mechanistic coupling, but then when C_3 is driven to extinction, the anti-phase coordination associated with the coupling of R_1 and R_2 (through competition) will return, again providing the opportunity for the invader to once again invade. The whole cycle than can repeat over and over, as illustrated in Figure 6.11b.

If we now turn to the phenomenological case (Figure 6.10a), the qualitative nature of the system is such that R_2 and R_1 have a net indirect positive effect on one another (on the principle that "the enemy of my enemy is my friend"). Thus, once the third resource species arrive, the two parent systems are coupled through that third resource species, but not in the competitive way shown in Figure 6.9d. Rather, the indirect positive coupling between R_1 and R_2 (by coupling through R_3) should cause the two underlying systems to become in-phase coordinated. There is no inherent reason why R_3 needs to be either in-phase or anti-phase coordinated with either R_1 or R_2, which leads to the possibility that the two main systems could be in-phase coordinated, thus leaving a space for the third species to invade. A particularly interesting case of this pattern is illustrated in Figure 6.12. The two independent subsystems are set in anti-phase motion to begin with, and then R_3 is introduced at about repetition 35 of the cycle. Coupling through R_3 causes in-phase coordination, as expected. However, the coupling also induces chaos in the two subsystems (Figure 6.12), yet, because of the coupling, the chaos is in-phase. This curious form, synchronized chaos, provides the invading species with windows of opportunity to invade, although the windows are completely unpredictable. R_3 effectively takes advantage of the periodic and simultaneous lowering of the predator density, thus permitting the species to enter the system (Figure 6.12). Thus, even though the two main competitors (R_1 and R_2) are very strong competitors against R_3 (that is the way the system was set up), all three coexist with the combination of the chaotic in-phase coordination of the dominants and the anti-phase coordination of the subdominant.

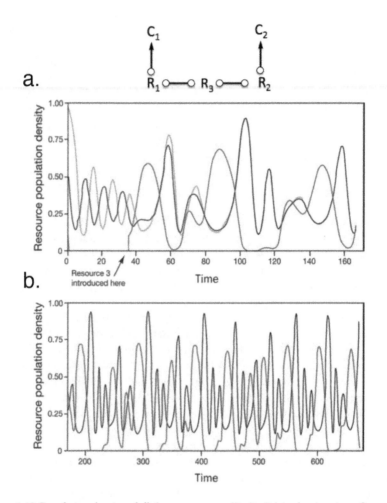

Figure 6.12 Population density of all three consumers (R_1, R_2, R_3) in the situation of resource competition. The darker shaded curves represent R_1 and R_2; the lighter shaded curve represents R_3. Note how the main oscillators begin anti-phase but rapidly are pushed into in-phase oscillations (this time chaotic), and how the subdominant competitor (R_3, lighter shade) is able to persist by effectively occupying the space when the other two competitors are at a low point.

Source: From Vandermeer, 2006.

Decomposition as an oscillatory process

Population processes operative at the level of soil are universally acknowledged as extremely important, as we elaborated in Chapter 2. Multiple species of bacteria act in ways to decompose organic material, catalize chemical reactions and eat one another, and fungi are known to be one of the most important decomposing organisms in nature. Viruses (called bacteriophages in the context of bacteria) act as major consumers in the system and competition takes on many different forms, some of which are rare or non-existent in the macro world. For example, bacteria, having no digestive system, actually digest their food outside of their bodies. The only way they can do that is to exude enzymes into their surrounding environment (called exoenzymes), let the enzymes break down larger cells, and then absorb the products of that chemical reaction. But if bacterium species A uses the energy to produce enzyme 1, what is to stop bacterium species B from saving energy by not producing enzyme 1, but rather cozying up to individuals of bacterium A and scavenging the products of the enzyme 1 that bacterium A produces? In a mathematical way, species B is a parasite on species A, creating, especially in a spatially extended environment, an arrangement whereby it eliminates B through "competition" (eating the food that A has made available locally), but then dying out itself (locally) because it has no food to eat. In addition to the obviously bacteriophage/bacteria and other forms of direct consumption, these complicated relationships that bacteria engage in are well-known from examples, and may very well be dominant forms of interactions in soils, suggesting not only that oscillations are likely, but that these oscillations are likely to be massively coupled.

In one study, for example, it was reasoned that the distance from a root tip back to its origin represents a profile in time, in that as the root grows, the bacterial communities associated with the root encounter the basic elements of the new environment created by the root itself. Thus, as the root begins growth, it begins exuding the carbohydrates on which bacteria locally grow. The bacteria grow and begin the process of competition or are attacked by predators or viruses (diseases of bacteria), in either case expected to show an oscillatory pattern. Results from a study that took bacterial samples along the main root of a wheat plant are shown in Figure 6.13. A wave-like pattern is clearly seen.

The oscillations of these bacterial populations are certainly suggestive of some sort of trophic connection, but what that connection might be is not clear. It could be, as suggested by the authors of the study, that natural enemies of the

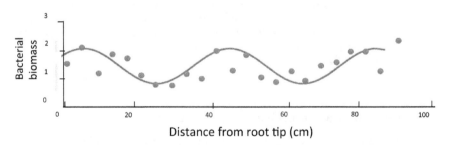

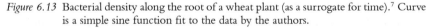

Figure 6.13 Bacterial density along the root of a wheat plant (as a surrogate for time).[7] Curve is a simple sine function fit to the data by the authors.

Source: Data extracted from Zelenev et al., 2000.

bacteria act to create predator/prey oscillations. However, it could be that the bacteria and their resources are simply acting as an oscillating system itself. Indeed, from the basic properties of plant nutrition (see Chapter 2), we might very well expect oscillatory behavior of decomposing bacteria, as outlined in Figure 6.14.

With this sort of seasonal cycle (Figure 6.14) we can expect oscillations between organic matter abundance and bacteria population, which could lead to yearly oscillations in the plant biomass production. A detailed study with wild rice in northern Minnesota suggests that just such an oscillation may be expected. As shown in Figure 6.15a, the mean seed weight per plant is related to the plant biomass, as would be expected. We can translate this relationship, approximately, into the mean seed weight this year into plant biomass next year, using a simple and arbitrary proportionality constant. Note that at lower plant biomass, the mean seed weight is much higher than at higher biomass. This sort of relationship then would lead to three basic groupings of points, as indicated

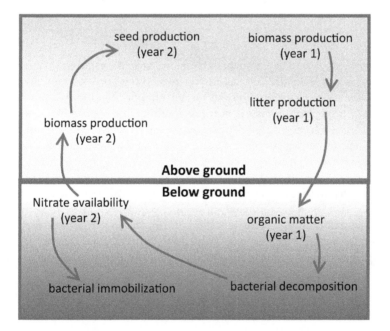

Figure 6.14 Generalized soil nitrogen dynamics and the production of above-ground biomass of a species of plant. Note how the availability of nitrate must be partitioned between plant biomass production (above ground) and bacteria "immobiliza-tion" (the nitrogen the bacteria must use for its own life processes), such that if organic matter is very high, the bacteria population will also be very high, which means that bacterial demand for nitrate will be high, thus reducing the amount available for plant biomass production, which, in turn, makes the biomass pro-duction the next year lower, and the organic matter lower, so the demand by the bacteria is lower, a repeating cycle.

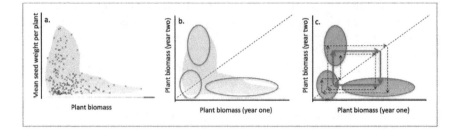

Figure 6.15 a. Relation between plant biomass and seed weight of wild rice (the more biomass, the more organic matter to fertilize the soil and feed the plants that then
produce more seeds).[8] b. Identifying three clusters of points. c. Inferred population dynamics of the wild rice population based on the three clusters of points
(qualitatively converting mean seed weight per plant into total plant biomass of
the next year, the ovals indicate rough areas within which the population will
reside each time period), illustrating the possible oscillatory dynamics that might
arise from this relationship. A yearly cycle suggested by the four bold arrows; a
5-year cycle indicated by the 10 dashed arrows.

in Figure 6.15b, low plant biomass in year one (*x*-axis) resulting in a group of
points located at low plant biomass in year two (*y*-axis), or low plant biomass in
year on resulting in a group of points located at high plant biomass in year two,
and, finally, high plant biomass in year one resulting in low plant biomass in year
two. Clearly such an arrangement suggests yearly oscillations, as illustrated in
Figure 6.15c (note that we suggest a yearly cycle and a more complicated cycle
involving repeated 5-year cycles).

   If it is true that such oscillations characterize bacterial populations in the soil,
what does that mean as we begin coupling those populations? The conventional
wisdom is that there are thousands of distinct populations of bacteria involved
in the essential ecological processes going on in the soil. Are these oscillatory
systems all coupled together, even if weakly? Do they synchronize? Do they
develop anti-phase coordination? Do they generate chaotic structures? All of
these questions await further research to be answered.

Seasonality and the Moran effect

We have introduced the issue of coupled oscillators as a problem of trophic connections (i.e., the various "consumers" and their "resources," as summarized in
Figure 6.2), where each trophic structure (predator/prey, host/parasite, etc.) is
an oscillator and they can each be connected to one another such that the two
are coupled. But there is a perhaps more obvious situation of coupled oscillators,
when an external force (such as seasonal structures of wet and dry or hot and
cold) acts on an oscillator. Almost all trophic structures are subject to some sort
of seasonal forcing. So, for example, when a pest species peaks in Finland in July

and completely disappears in December, no one would suggest that this implies its "natural frequency" is 12 months.

A convenient modification of the simple model of coupled pendula might be useful here. Suppose we attach a fixed rod to one of the pendula and attach the other end of the fixed rod to an electric motor, such that the pendulum is completely driven by the electric motor (as in Figure 6.16). The second pendulum will be, to some extent, driven by the first pendulum, but itself have only a trivial effect on the first pendulum. In other words, pendulum 1 is a "driving" oscillator, dependent only on the driving force of the electric motor. Pendulum 2 is "forced" by pendulum 1, but the spring connecting pendulum 1 with pendulum 2 is flexible, with the degree of flexibility, as before, corresponding to the degree of coupling of the two oscillators. The difference between the coupled oscillators of Figure 6.4 and the forced oscillator of Figure 6.16 is that the forced situation has only a trivial amount of feedback from the second pendulum onto the first. The most obvious ecological application of this arrangement is the wet and dry (or cold and hot) season change representing the forcing oscillator (the pendulum driven by the electric motor in Figure 6.16), driving the predator prey oscillator.

Some fairly obvious patterns can be anticipated from the basic arrangement of the forced oscillator system. If, for example, the natural frequency of the consumer/resource system is equal to the forcing frequency, the ratio between the forcing frequency and the natural frequency will be 1.0, and the consumer/resource system will oscillate just as it did before being forced, perhaps more strongly (with higher amplitude), but with the same frequency. If, to take another elementary example, the forcing frequency is 1/12 the natural frequency, the ratio between the natural and forcing frequencies will be 12. As before, we expect the consumer/resource oscillations to be reinforced by the forcing oscillator, but this time the consumer/resource system will oscillate 12 times for every oscillation of the forcing system. So, for example, suppose the consumer/resource system is a small insect pest and a disease that attacks it regularly, such that the disease peaks about once per month. The seasonality (think of wet/dry season) influences the system in that the rate of reproduction of the insect is slightly higher in the wet season than in the dry season, but that is about all. So, if there were no seasons at

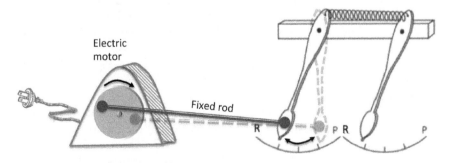

Figure 6.16 Physical model of the idea of a "forced" oscillator

all, the insect population would oscillate with a frequency of one oscillation per month. Adding the forcing would not really change that much, since the forcing itself is a 1-year cycle, so that the insect population simply has its oscillation frequency "encouraged" by the seasonal oscillations.

But we can imagine an alternative scenario. Suppose that the insect/disease system in a perfectly homogeneous environment oscillates not quite once per month. Suppose it oscillates 117 times every 10 years, or 11.7 times per year (that is, its "natural" cycle is 11.7 per year). What, then, would be the effect of releasing the system into a seasonal (cycling every 12 months) environment? The answer to that question can get a bit complicated, and is to some extent dependent on how sensitive the population actually is to the seasonal forcing. It is as if the "fixed rod" in Figure 6.16 is flexible and has a definite forcing effect on the insect population, but not necessarily all that strong. But there is a bit of a generalization that can be seen fairly easily. The population is likely to come into some sort of coherent relationship with the seasonal forcing. That is, rather than oscillating 11.7 times per year (117 times every 10 years), the population will be "coaxed" into 12 times per year cycle by the influence of the seasonal forcing. The seasonal cycle will effectively continuously reset the natural oscillations so they are coherent with the seasonal cycles.

Understanding this coherence can get quite arduous, with a fair amount of very elaborate literature on the subject.[9] The basic conclusion is that a force operating on an oscillator (e.g., seasons operating on a consumer/resource system) will result in the oscillations of the system to form a rational (i.e., integer number divided by another integer number) frequency related to the seasons. Sometimes this number will be small (e.g., two cycles per season, three cycles per season), but sometimes it can be rather complicated. Imagine, for example, an insect population that cycles once every 3 months based on its natural frequency, that is, its oscillatory frequency if it were theoretically in a homogeneous, non-seasonal environment would be four cycles per year, each cycle 3 months long. We illustrate such a population in Figure 6.17a. In central Mexico the rainy and dry season is on a 12-month cycle, more or less 6 months of dry season followed by 6 months of wet season. So if the insect is affected by rainfall, its 3-month cycle will be nicely calibrated with the 6-month rainy season and its cycles with the seasonal forcing will be modified, higher than normal populations in the rainy season, lower than normal populations in the dry season (Figure 6.17b). However, if the seasons are very strong, if the dry season kills virtually all the insects, the natural oscillations will be affected more strongly, and the population will tend to oscillate more according to the specific seasonal dictates (Figure 6.17c). The population will tend to oscillate twice per year (Figure 6.17c). At a different seasonal extreme, in northern Brazil the rainy and dry season alternate on a 3-month cycle. Again the insect will be nicely calibrated to cycle according to the season, but now the seasonal cycles will have peaks that always coincide with the peaks of the natural population cycles, and the resulting population cycles will be once per season (or four times per year; Figure 6.17e).

But what about some place intermediate between Mexico and Brazil, where the seasonality is more complicated, with a rainy season of 4 months, followed by a dry season of 2 months, followed by a second rainy season of 2 months

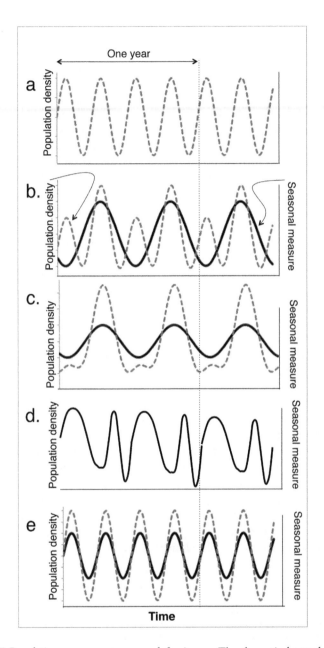

Figure 6.17 Population responses to seasonal forcing. a. The theoretical population oscillations with no forcing – obviously four cycles per year. b. A space in which
seasonal forcing is twice per year (black solid curve) and the resulting forced
population remains at four cycles per year, but with one of the peaks reduced by
the dry season. c. More extreme forcing than in (b), such that the lower population peak effectively disappears and the population cycles along with the forced
oscillator (the seasons). d. Area intermediate between two seasons per year and
four seasons per year, illustrating the complicated forcing patterns that may exist
in a forcing function even as simple as the seasons. e. Result of seasonal forcing
that corresponds in frequency with the natural frequency of the population.

followed by a dry season of 4 months (Figure 6.17d)? If the insect is not capable of surviving the dry season, we see that its oscillations can be completed within the 4-month rainy season, but the second rainy season will not allow the natural oscillations to be completed. So the oscillations might be forced into a once per year oscillatory pattern, effectively determined by the seasonal forcing. The resulting population oscillations can get very complicated in this intermediate zone, despite the fact that their response to forcing at the extremes is obvious.

When a variable (say, an insect pest population, or the amount of nitrate in the soil) is oscillatory and the oscillations are completely driven by an external force, the phenomenon is known as the Moran effect. Stimulated by the apparent cohesion of population oscillations of the Canadian lynx across all of Canada, Australian statistician Pat Moran introduced the idea with regard to independent populations in 1953. The basic idea is one of a forcing oscillator (like the electric motor in Figure 6.16) acting simultaneously on two or more populations, thus resulting in the synchronization of both populations. The lynx population is oscillatory on approximately a 10-year cycle, all across Canada, and the oscillations are synchronized with one another. Moran suggested that the synchronization was determined by the basic meteorological fact of the seasons oscillating between winter and summer, a hypothesis that seems to be well-corroborated through more recent work.[10]

While not exactly equivalent to the normal use of the Moran effect, we think it is useful to think of the basic question as similar to the idea of fundamental versus realized niche, or the equivalent idea of exogenous forces versus endogenous forces causing pattern. Here we are talking about the "pattern" called oscillations or the pattern called synchronized oscillations. So, reverting to the simple physical model, of Figures 6.4 and 6.16, when we see a population oscillating, we can ask, are those oscillations caused by internal (endogenous) or external (exogenous) forces? Any oscillation (e.g., a predator/prey oscillation or a substrate/bacterial population) will have a "natural" frequency, the frequency caused by the internal operation of the oscillator. This is the oscillation that arises from the simple fact that we are dealing with an elementary trophic structure (Figure 6.2). However, it is entirely possible that oscillations are determined, not by the natural interaction of trophic elements, but by an external force, such as seasonality. The parallel with the older ideas of fundamental versus realized niche is, we think, obvious.

Notes

1 Bennet et al., 2002 (Huygens quote).
2 Much of this analysis was anticipated long ago by Arthur Winfree in his classic work (1980) and more recently in a popular form by Strogatz (2003).
3 Vandermeer, 1993, 2004a.
4 An excellent example can be found in Benincà et al., 2009.
5 Armstrong and McGehee, 1980; Levins, 1979.
6 Details of this analysis can be found in Ong and Vandermeer, 2015.
7 Data are from Zelenev et al., 2000. Important follow-up studies are presented in Zelenev et al., 2005.
8 Walker et al., 2010.
9 An important early paper on this topic is King and Schaffer, 1999.
10 Blasius et al., 1999.

7 Multidimensionality

The most fundamental aspects of ecology are conceptualized in a very low dimensional space, even though it is understood that ecology in the real world involves many actors, which is to say, it is multidimensional. As mentioned previously, the simple statement that ecosystems are complicated (e.g., multidimensional) sometimes obfuscates rather than clarifies, and the admonition that reality is complex is sometimes nothing more than an excuse to not think carefully.

An antidote to the temptation to throw up our hands and say "it's just too complicated to think about" can be found in thinking about the dimensionality question in a historical framework. One- and two-dimensional systems certainly have dominated our thinking, largely because they are easiest to visualize. Yet, some of the most important conceptual ideas in ecology are evidently multidimensional. In the past we have dealt with this problem in various ways. For example, the equilibrium theory of island biogeography[1] effectively takes a very large number of species into account. However, instead of considering each species individually, it examines the dynamics of the movement of species and the resulting species richness at equilibrium. Likewise, the theory of a metapopulation[2] is about an effectively infinite number of individuals, but rather than taking each individual into account it examines the proportion of patches that are occupied by the species in question. These are examples of mean field models, and the main characteristic of these models is that they replace many interactions with their means (thus the appellation "mean field") and therefore reduce the dimensionality of the system significantly. Another approach is to effectively take the organisms out of the picture entirely and examine only processes. So, for example in ecosystem ecology trophic dynamics, including nutrient cycling, and energy flow, are implicitly multidimensional systems, but by examining the flow of energy and the cycling of matter the multidimensionality conundrum is consequently avoided. Many other examples could be cited.

Thus, we see that ecological conceptualizations tend to be concentrated either at the level of one or two dimensions (e.g., competition theory, host/parasite interactions) or, with a mean field style approach, at the level of an effectively infinite dimensionality (e.g., rate of arrival of new species, population extinction levels, nutrient flow rates, etc.). Although certainly much understanding has emerged from all of these approaches, real ecological systems, especially agroecosystems, actually operate at an intermediate dimensionality. Thus, there needs to be some attention to the "more than two" but less than "infinite" levels.[3] It is, thus, justifiable to consider dimensionalities that are larger than two, but not so large that we effectively have a "mean field" approach. And the first step in this direction is adding the third dimension. Indeed, the outcome of some processes or ecological interactions can be changed dramatically by simply going from two to three dimensions, a point to which we now turn.

Three-dimensional systems

There is something very special about continuous two-dimensional systems. Recall the idea of analyzing things in phase space by looking at the vector field (as discussed in Chapter 2). For any continuous model in two dimensions, if you examine the vector field, you see that each and every point in the space has one and only one vector associated with it. This means that trajectories cannot cross one another, as illustrated in Figure 7.1.

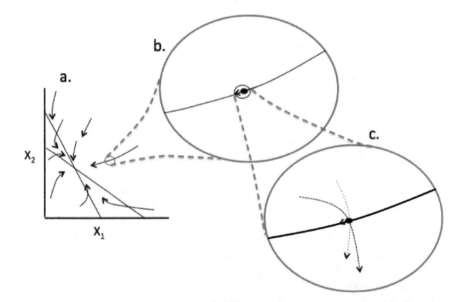

Figure 7.1 Illustration of the impossibility of trajectories crossing one another in two-dimensional space. a. Basic model (e.g., Lotka–Volterra interspecific competition) showing multiple possible trajectories (the arrow at the end could represent the vector in the vector field associated with the point at the end of each trajectory). b. expansion of the small area around one of the points on one of the trajectories, showing the vector associated with that point (the arrowhead) along with a segment of the original trajectory. c. Even more microscopic expansion of the small area around the point, illustrating how two alternative trajectories could not cross that point. There is by definition only one vector associated with each point in the phase space, so the alternative vectors implied by the trajectories indicated by the dotted arrows would be impossible. Only a single trajectory can pass through any given point in the space.

It is clear from Figure 7.1 that, because every point in the space can have only one vector, there can be only one trajectory that passes through that point and thus trajectories can never cross one another. This is an extremely important point. For example, the situation in Figure 7.2a is not allowed, but the situation in Figure 7.2b, which shows a three-dimensional space, is perfectly possible. Obviously this places strong restrictions on what the system is allowed to do over the long haul in a two-dimensional space. Recall, for example, the various chaotic attractors from Chapter 4 (e.g., Figure 4.3b). This chaotic attractor (Figure 4.3b) occurs in a three dimensional space but we illustrated it in only two dimensions. If the system was really a two-dimensional system, the way the trajectories cross would not possible. We illustrate this idea in Figure 7.2, showing how the addition of another dimension, even if a very restricted range dimension, solves this problem (i.e., trajectories apparently can cross, but only because there is actually another dimension; Figure 7.2c).

The type of restriction illustrated by this example actually is representative of a general problem. Restricting a system to two dimensions dramatically limits the potential behaviors it can represent. For example, chaos is not possible at all in a continuous two-dimensional system. Adding that third dimension, even if you restrict its range to an effectively infinitesimal amount, opens up an unbounded variety of behaviors that could be produced. In other words, a major importance of multidimensionality is that moving from two to three dimensions can generate a qualitative change in the system's behavior.

Another recently appreciated framework that emerges in three or more dimensions is the possibility of an intransitive loop. This is most easily seen in the process of interspecific competition, although the phenomenon is seen in all sorts of multidimensional applications.[4] Historically we have long been

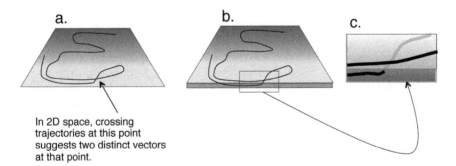

a.

b.

c.

In 2D space, crossing trajectories at this point suggests two distinct vectors at that point.

Figure 7.2 Illustration of how a three-dimensional system permits trajectories to appear to cross when viewed in two dimensions, even if the third dimension is very small. a. A trajectory in two dimensions appears to cross itself – an impossibility since only a single vector can be associated with each point. b. Adding a third dimension to the system enables the two trajectories to seem like they cross. c. A close-up of how the apparent point of crossing is really separated in the third dimension.

accustomed to the fundamental four possible solutions to the basic Lotka-Volterra approach:[5] (1) species A always beats species B, (2) species B always beats species A, (3) the two species coexist in perpetuity, and (4) one or the other species wins, depending on where the system started. It may seem obvious that with three species a similar arrangement could be true, and to some extent that is true: there are actually 11 qualitatively distinct possibilities. But one arrangement has been recognized as especially important. Let us assume that we have three species, A, B, and C, in competition with one another, and that the competitive outcome is always fixed when any two of them are competing alone, which is to say, no two species can coexist in perpetuity. This might lead us to believe that if we put all three species in the mix, only one would survive. However, it is possible that the three form an "intransitive loop," an arrangement similar to the familiar rock-scissors-paper game, A beats B, B beats C, and C beats A. Note how this basic arrangement violates the standard mathematics lesson of transitivity from elementary school, that is, if $A > B$, and $B > C$, it must be the case, by the transitivity property, that $A > C$. But in the rock-scissors-paper game, $C > A$, which violates that property, and thus we call it an intransitive (or non-transitive) loop. So, if we have three species in the system and it is the case that any pair of them cannot coexist, due to competition, it is also the case that all three *could* coexist if living together, as long as they formed an intransitive loop.

There seems to be a tacit assumption when thinking of competitive interactions. For example, in the practice of intercropping, when crops are generally of the same stature and generally share the same ecological requirements (i.e., have similar niches), any pair of crops will not be likely to overyield (i.e., interspecific competition will be strong and it would be better to grow them as separate monocultures).[6] However, if there are three crops forming an intransitive loop, it is possible that the three together will indeed overyield, even though none of the three pairs would do so. In the plant competition literature more generally there is a very similar assumption, frequently unstated, that plants of generally the same stature and requirements will form a strict competitive hierarchy (i.e., species A beats species B beats species C beats species D, etc.). Although it may seem intuitively obvious, the assumption is not necessarily true. An example is offered by Daniel, a Nicaraguan corn farmer of our acquaintance. As his corn is growing, a relatively dense growth of morning glory vines begins to grow beneath the corn stalks (Figure 7.3a). Daniel does not "weed" them. Also in the general area grows purple nutsedge (Figure 7.3b), another weed that he knows to be highly aggressive against the corn plant. But the morning glory vines grow over the purple nutsedge and do not allow it to prosper (Figure 7.3a). After he harvests his corn, Daniel lets the morning glories grow as much as they want, effectively completely covering the sedge and any other weeds that might try to get established. Then, when planting time comes, he burns off the morning glories (leaving their rootstalks for the next planting) and plants the corn. So, effectively there are three agents here: the sedge, the morning glories, and the corn farmer and his corn. And they obviously form an intransitive loop since in the end the "corn-farmer-and-his-corn" beat the "morning glories," while

Figure 7.3 Two typical weeds in many tropical agroecosystems. a. Morning glory vines (*Ipomea purpurea*) covering a host of other potential weeds – sedges are long gone due to the intense shade cast by the morning glory. b. Morning glory vines penetrating a clump of purple nutsedge. c. The highly aggressive sedge, purple nutsedge (*Cyperus rotundus*), perhaps the most feared weed in traditional maize production in this region of Nicaragua.

Source: Authors.

the "morning glories" beat the "sedge," and the "sedge," if given the chance, will beat the "corn." Note that all three agents are, in principle, persistent in the system.

A concrete example of how an intransitive loop can operate is offered by the conundrum of the potential of escape of the particular form of GMO crop that confers resistance to caterpillar pests, as previously discussed in the context of chaos in Chapter 4 (Figure 4.11). Another intriguing theoretical example is offered by an evolutionary consideration formulated in a geometrical space where three species are involved in a competitive intransitive loop. Authors Frean and Abraham (2001)[7] note that if one of the species has two genetic varieties, one a strong competitor and the other a weak competitor, it is conceivable that the strong competitor will be eliminated from the system, thus the surprising title of their paper, "The Survival of the Weakest". We can imagine various relevant scenarios with, for example, weed communities. We already know that plant communities frequently contain intransitive loops within their overall species diversity. Indeed, the expectation, from a group of plants that compete with one another strongly, is that if you randomly sample them three at a time, about 25% of the time you will come up with an intransitive loop.[8] What that might mean for eventually being able to manage weeds depends on our understanding of how intransitive loops operate.

We have already seen an interesting application of intransitive loops, emerging from a combination of a simple trophic structure, such as consumer eats resource, or predator eats prey. In Figure 3.5 we diagrammed the basic process of a predator/prey or consumer/resource system distributed in a spatial context. The cycle, empty space gives rise to prey population, which gives rise to predator/prey combination, which gives rise to empty space, is, formally, an intransitive loop. As we discussed in Chapter 3, this intransitive loop can be either stable or unstable, depending on the three rates: rate of migration of prey, rate of migration of predator, and feeding rate of predator. This fundamental way of thinking of trophic structures distributed in space is currently the subject of ongoing research.[9]

Multidimensional systems

The modern industrial agricultural system is not normal. That is to say, taking the past 10,000 years of agriculture and the entire globe into account, for the most part agroecosystems have been multidimensional in several senses. Usually many species are involved, either in the planned nature of the farm (what we refer to as the "planned biodiversity") or in the inevitable arrival of many species that associate themselves in some way with the system (the "associated biodiversity"), or in the many and varied functional forms ("multifunctional farming systems"). The *dehesa* system of the Iberian Peninsula is an excellent example (Figure 7.4). Covering about 3.5 to 4 million hectares in the Iberian Peninsula (Spain and Portugal), it has been in existence as an agroecosystem since the Neolithic period, and probably derives from an original savannah system, subject to periodic fires.[10] As an agroecosystem, the major planned plant

Figure 7.4 An example of the *dehesa* system in Bollullos Par del Condado, Huelva, southern
    Spain. Note the harvested bark on the cork oak tree to the left.

Source: Wikimedia Commons: https://commons.wikimedia.org/wiki/File:Dehesa_Boyal._Bollullos _Par_del_Condado_(Huelva).jpg.

biodiversity consists of two main species of oak, one of which produces cork (effectively an industrial product) and one of which produces mainly acorns. Beneath the scattered oaks is grassland/pasture, sometimes managed intensively and sometimes not. Grazing on the grassland are goats, sheep, pigs and cattle. One of the high-end products is the famous *"pata negra"* ham, an extremely high-value commodity that is produced from the pigs who feed mainly on the acorns. Included in more open areas is the production of various grains. In years of low grazing pressure, sometimes the woody vegetation begins to take over the grasslands, requiring the hiring of local labor for hand clearing. Thus, from a strictly biodiversity point of view, the planned biodiversity includes three major plant types (two oaks and grasses), several grazing animals, and planted grains, while the associated biodiversity includes the woody vegetation that occasionally takes over locally, plus all the potential herbivores that attack the oaks and grasses and grains, plus all the ectoparasites that attack the grazing animals, as well as hundreds, perhaps thousands of species of insects that inhabit the *dehesas*. The management aspects are obvious in several ways, managing the trees (minimal effort), clearing woody vegetation, plowing for grain production, harvesting grains. In short, it is a system, a very old system, that is multidimensional both in biodiversity and in the diversity of agricultural activities.

The first significant foray in theoretical ecology beyond three dimensions was by Richard Levins in 1968. Levins basically expanded the idea of going from two species (which by then was a well-watered theme in ecology) to three, and expanded to a very large number. The idea he spawned was the "community matrix," a major conceptual breakthrough in population and community ecology.[11]

In its original conceptualization, the community matrix is a table of interspecific competitive effects, also referred to as competition coefficients (later formulations include other types of interaction coefficients, predator/prey, mutualisms, etc.). The competition coefficient is the reduction in growth rate of the target species per individual of an effector species (interspecific competition), relative to the reduction in growth rate of the target species per individual of the same species (intraspecific competition). So, for example, if adding a single individual of perennial ryegrass (*Lolium perenne*) reduces the growth of a population of its own species by 10%, but adding a single individual of white clover (*Trifolium repens*) reduces the growth rate of the perennial ryegrass population by 5%, the competitive effect, or competition coefficient, reflecting the effect of white clover on bent grass is $0.05/0.10 = 0.5$ (Figure 7.5).

The columns of the community matrix represent the effect that one species has on another, the rows represent the response of a species to the effects of all the other species. An example is given in Figure 7.6a for seven species of herbs in North America (mostly European invasors), where, for example, the effect of one individual of perennial ryegrass (*Lolium perenne*) on white clover (*Trifolium repens*), relative to the effect of white clover on itself is 2.70. With reference to Figure 7.6a we can say a certain number of things about what we expect from this weed community based on the community matrix. For example, the sum of

Figure 7.5 Exemplary weedy plant species included in the community matrix in Figure 7.6. a. *Trifolium repens,* the white clover (source: Forest and Kim Starr https://commons .wikimedia.org/wiki/File:Starr_070313-5645_Trifolium_repens.jpg). b. *Lolium perenne,* the perennial ryegrass (source: Rasbak https://commons.wikimedia.org /wiki/File:Engels_raaigras_(Lolium_perenne).jpg). c. *Amaranthus retroflexus* as a major weed in maize production (source: Phil Westra, Colorado State University, Bugwood.org).

a

Species	Lp	Tp	Tr	Rc	Ca	Ar	Pp	Row sum
Lolium perenne	1.00	0.03	0.09	0.08	0.05	0.10	0.14	1.27
Trifolium pratense	6.20	1.00	3.05	1.25	0.95	1.00	1.40	14.85
T. repens	2.70	1.39	1.00	0.30	0.57	0.74	0.44	7.15
Rumex crispus	10.06	5.26	1.60	1.00	2.15	2.53	1.57	24.77
Chenopodium album	8.94	4.01	1.63	0.43	1.00	1.60	0.72	18.33
Amaranthus retroflexus	9.61	1.13	0.51	0.47	1.15	1.00	0.63	14.50
Phleum praatense	3.16	0.69	0.26	1.02	0.95	0.59	1.00	7.68
Column sum	42.27	13.46	8.13	4.39	6.83	7.57	5.90	88.54

b

Species	Lp	Tp	Tr	Rc	Ca	Ar	Pp
Lolium perenne	1.00						
Trifolium pratense	1.00	1.00	1.00				1.00
T. repens	1.00		1.00			1.00	1.00
Rumex crispus	1.00	1.00	1.00	1.00	1.00	1.00	1.00
Chenopodium album	1.00	1.00	1.00		1.00	1.00	
Amaranthus retroflexus	1.00	1.00				1.00	
Phleum praatense	1.00				1.00		1.00

Figure 7.6 a. Estimated competition coefficients placed in a "community matrix." Relative rates of competition calculated from data in Goldberg and Landa, 1991. (Note that a negative value represents a facilitative effect.) b. Competitive outcomes matrix from (a).

the rows of the matrix indicates the effect of all the species in the community as a whole on the species in that row, basically indicating which species in the community that particular species must respond to. In this sense the row sum is frequently taken to be inversely proportional to the ability of the species to *respond* to the competitive effects of the rest of the species (i.e., to resist competition from all other species). For example, in the community matrix presented in Figure 7.6a, curly dock (*Rumex crispus*), is a very bad response competitor in this community since the sum of competitive effects of all the species on curly dock is 24.77, the largest sum of all. Another feature of the community matrix is the column sum. This represents the total *effect* that a particular species has on all the other species in the community. Again, curly dock is the poorest of all the competitors in terms of its effect on the other species in the community (overall effect on all the other species is 4.39). The row sum is frequently taken to be an inverse measure of the "response" competitive ability, while the column sum is a measure of the "effect" competition. The balance between effect and response is sometimes a useful way to characterize a community. So, for example, with the information in Figure 7.6a we could probably predict that the perennial ryegrass (*Lolium perenne*) would take over from the other species since it is a very good effect competitor (column sum of 42.27) and also a very good response competitor (row sum of 1.27). Not surprisingly, perennial ryegrass is a notorious indicator of species–poor meadows in its native European range.

Sometimes it is useful to transform the community matrix into a "competitive outcomes matrix." That is, if we presume that all pairwise interactions are completely hierarchical, in the sense that only one species can survive the competitive interaction, we can consider every pair of competition coefficients and replace the highest one with a 1 and the other with a 0. So, for example in Figure 7.6a, Pp has a competitive effect of 0.14 on Lp, whereas Lp has a competitive effect of 3.16 on Pp, so it is likely that at least in a very small region (say a flowerpot), LP will always prevail. So the effect of Lp on Pp is 1 and the effect of Pp on Lp is 0 (Figure 7.6b). That matrix of 0s and 1s is the competitive outcomes matrix. It is frequently presented as a graph with arrows pointing to the loser in any pairwise competition.

Using the information in the competitive outcomes matrix, we can construct a competitive outcomes network using arrows to indicate who wins over whom in competition (Figure 7.7a). A quick glance at the graph immediately reflects what we noted above, that the perennial ryegrass (*Lolium perenne*) beats all other six species in the system. And, as we noted, in Europe it is regarded as an "indicator" of a low diversity meadow (having competitively excluded everything else). We can suppose that a program of eliminating this noxious weed has been successful, which leads us to the graph in Figure 7.7b. An obvious feature of the new graph is that Rc loses competitively to all five other species. Thus we can expect that species to be eliminated, with the result that the remaining community is composed of five species, as represented in Figure 7.7c. And here we see that there is no other real obvious outcome; no species that will obviously exclude all the others nor one that will obviously be excluded. Furthermore, as

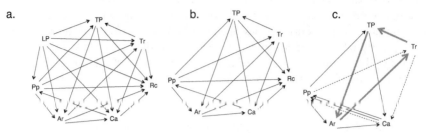

Figure 7.7 a. The competitive outcomes network of the data in Figure 7.6b, based on the competitive outcomes matrix (1 = competitively superior; 0 = competitively inferior); b. The resulting graph if Lp is eliminated; c. The resulting graph after Rc is eliminated due to competition. Note that the resulting 5-species network includes four 3-species intransitive loops (indicated with different colors, patterns and thickness of line).

we can see, there are four intransitive loops that characterize the system (coded in color and dashed and various bold lines to indicate which is which). Note that Pp is involved in all three of the loops, so we might expect something dramatic to happen if this species were artificially removed. Indeed, if Pp were removed, all other species would be negative against Ca, suggesting its eventual competitive exclusion, resulting in a single intransitive loop of Tp, Ar, Tr. On the other hand, if the four-loop system (Figure 7.7c) was allowed to continue, over time, it would probably result in the elimination of Ca (since it is the only species that is negatively affected by three of the four other species), which would lead to the complete lack of competitive control over Pp, allowing it to eliminate all three other species. The point of these speculations is that the underlying structure of this weed community depends critically on individual exclusion events and their order of occurrence and is largely unpredictable, over the long term, knowing only who competes with whom.

The promise of food webs

The spaghetti network we introduced earlier (Figure 1.1 in Chapter 1) remains a useful call to humility, but offers little in the way of organizing the world in such a way that "understanding" emerges. It was in 1947 that Raymond Lindeman suggested organizing complicated systems of energy transfer in terms of trophic structure (effectively who eats whom) and asking questions of nutrient cycling and energy transfer pathways as the foundation for understanding. This approach becomes quite complicated, approaching the spaghetti graph, as more elements are added, and as the ecosystem is viewed through a finer and finer lens. Recently a new focus has been applied to these multidimensional systems, network theory. In the previous section, with the competitive network we effectively introduced the idea of a network.

Network theory, as a general subject, begins by asking the question, how much can we know about a system if all we know is which element is connected to which other element? Applied to anything from the internet to electrical distributional systems, its most obvious biological application is to disease. Suppose, for example, a farm has banana trees scattered in the periphery of an agricultural field, in a form suggested in Figure 7.8a. Now suppose that we wish to represent a banana disease that is transmitted by local contact or very local wind currents such that only nearby trees can infect one another. We can do this by constructing a basic network for this disease, as in Figure 7.8b. Note that if the plant indicated by the small arrow becomes infected, it sequentially infects all nearest neighbors. In this artificial example each plant has at most two neighbors, and infection is at 100% after 11 weeks (assuming it takes about a week for one plant to infect its neighbor).

If we now imagine that a path had been worn such that people regularly walked through the farm, and that the human traffic carried propagules of that disease. In that case, there would be another connection, one that is determined by the path, as suggested in Figure 7.8c. This arrangement is frequently referred

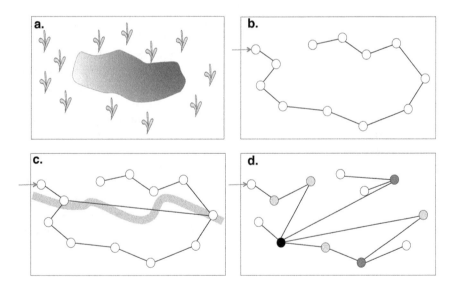

Figure 7.8 Idealized farm with a central region of some crop and banana plants in the periphery. a. Diagram of the central cropping area and the individual banana plants. b. Representation in a graph where the connection is based only on the nearest proximity of plants. c. Illustration of how a path through the farm can make a connection that would likely not have existed without the path, generating, in principle, a small-world network. d. Illustration of how preferential attention to individual banana plants can lead to a scale-free network.

to as a "small-world network" based on the idea of moving rapidly to different parts of a network through just a few links.[12] The idea is related to the famous experiment done by Jeffrey Traves and Stanley Milgram in 1967 in which they selected individuals in in Nebraska and Boston and asked them to generate an acquaintance chain to a specific person in Massachusetts. The results were surprising. The average number of intermediaries was 5.2, giving rise to the popular idea of six degrees of separation, the idea that people, and perhaps all things in the world for that matter, are separated from each other by a maximum of six steps. For many purposes in ecology the small-world structure generates remarkable results, especially in disease ecology, but also in many cases where space is involved. For example, in the regular graph of Figure 7.8b it took 11 weeks for a complete epidemic, horribly slow compared to the 7 weeks it would take in the small-world network (Figure 7.8c). A small-world network is almost always far more complicated than the graph pictured in Figure 7.8c, which was constructed only for heuristic purposes, a point to which we return momentarily.

A completely different kind of graph may represent certain biological systems. Imagine, for example, that a kind of bird perches regularly on the banana trees. But the birds for some reason may prefer some trees to others. Thus, in Figure 7.8d the nodes are shaded in proportion to the preference the birds have for them: the darker the color, the greater the preference. This sort of graph construction is known as "preferential attachment" and results in what is known as a "scale-free" network. If you take the number of connections each node in a graph has, the distribution of those connections is formally known as the "degree distribution" and represents a sort of measure of how preferred a node is, if indeed the network was constructed with the preferential attachment prejudice. Remarkably, if you plot the log of the number of attachments per node against the log of the frequency of occurrence of that number, a straight line emerges. Recall from Chapter 3 that a straight line on a log/log graph represents a power function and reflects the fact that no matter what scale you consider (no matter what range on the x-axis), if you fit a line to the points within that range you get the same slope of the line. In other words, the process described by the graph is independent of the scale that is used, it is "scale free". For example, the network in Figure 7.8d has one node with five connections, two nodes with three connections, four nodes with two connections and five nodes with one connection. We leave it to an exercise for the reader to plot the logs of 1, 2, 4, and 5 against the logs of 5, 3, 2, and 1 to verify that it is approximately a straight line.

We now return to the idea of a small-world network. In Figure 7.9 we illustrate a process of transformation of a network from regular to completely random. Take the network on the extreme left, the one we constructed with the artificial banana example, which represents a regular network, and randomly eliminate one connection. Then rewire one of the nodes that was connected to a random node, creating a small-world structure (Figure 7.9b), which is more or less what we did in Figure 7.8c. Now, rewire another, and another, and another, and another, to get the network in Figure 7.9c, another small-world structure. Now continue rewiring until all the connections have been randomly assigned,

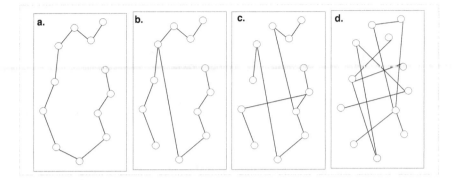

Figure 7.9 The transformation of a network from regular to random. a. A regular network (as in Figure 7.7b). b. A small–world network. c. Another, more randomish, small-world network. d. A random network.

as in Figure 7.8d. So we begin with a regular network (Figure 7.9a), proceed through a series of small-world networks (Figure 7.9b and 7.9c), until we arrive at a random network (Figure 7.9d). It is clear that the extremes, regular and random, embrace a whole series of small-world networks. A great deal of literature has evolved asking the question whether a particular network is small world, and if so, to what degree.[13]

The example given here of a plant disease spreading through a population of banana trees is quite a straightforward example of the use and utility of network theory. We could easily deduce that the rate of spread of the disease was strongly affected by the structure of the network, and the insight of small-world network construction generated the qualitative understanding that diseases spread through small-world networks much faster than through regular networks. There are, however, many applications that are less obviously legitimate, some of which could actually contribute to obfuscation rather than understanding.

Consider, for example, a food web. There is much interest in representing food webs as networks, with the nodes being species and the edges (connections) representing energy flow. But we can hardly even begin the discussion without acknowledging that much of what is "beautiful" about network theory (e.g., small-world networks, scale-free networks) becomes immediately irrelevant, or at least more difficult to apply in a fashion that generates understanding. For instance, the two alternative trophic structures illustrated in the previous chapter (Figure 6.10b), if represented as a simple network graph, would look like the graph in Figure 7.10a. Knowing, as we do, that the system is one of coupled oscillators (Figure 6.10b), and having concluded that the oscillatory nature is a key element of the system, a point to which we seek to focus our analysis, representing it as a simple network graph (as in Figure 7.10a) provides little understanding, perhaps a bit of obfuscation instead. At a minimum, we need to indicate which direction the energy is flowing, as in the network of

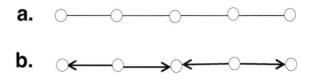

Figure 7.10 Representation of the coupled oscillator system presented in Chapter 6 (Figure 6.10b) as a network graph. a. A simple network. b. A directed network.

Figure 7.10b. A network representation of this sort is referred to as a directed graph, and some of the elegant theory already developed for simple graphs ceases to be applicable. To ask the question, "how much can we say from a simple knowledge of what is connected to what (without stipulating direction) for food webs?" even on cursory examination would attract the answer, "not much." Indeed, a significant research question emerges quite naturally, although not yet investigated: what is the main determiner of the dynamics of a food web, the undirected network structure or the directions themselves? Can we, to put it differently, replace the values in the community matrix with binaries (1 = some effect, 0 = no effect), and retain the ability to discern the system dynamics? As we discussed in the previous section, certainly some qualitative speculations are aided by such a procedure, but much is also left out.

A major thrust of network analysis has been centered on the question of resilience of an ecological system. At its most fundamental level, the question is "how does the topological structure of a network provide resilience to a system?" And one of the most common procedures is to specify, to as great a degree as possible, the structure of a network, then systematically remove species and ask what happens to the network structure as a whole. A simple artificial example is presented in Figure 7.11. If one of the carnivores is eliminated from the system (e.g., Figure 7.11b), knowing the simple network structure can only tell us that nothing dramatic is expected to happen. While that carnivore could, conceivably, control herbivore 1, which is a pest of the crop, and eliminating it (the carnivore) could conceivably result in herbivore 1 exploding and eliminating the crop entirely, such an outcome is not discernable from our representation of the system as a simple network graph. As a matter of fact, eliminating either of the carnivores from the system would result in nothing predictable happening to the system, based on the simple network representation. However, if our elimination were to target the crop itself (i.e., abandon agricultural production), some major changes would happen to the system. In particular (as is evident in Figure 7.11c), the entire subnetwork connected to the crop would be eliminated. Here, with additional knowledge that two of the nodes represent primary producers (the crop and weed are both plants and provide the energy base to the system), the simple network diagram indeed presents us with a certain degree of predictability.

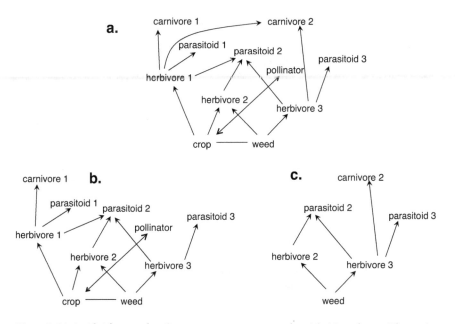

Figure 7.11 Artificial example of an agroecosystem network, with 11 nodes. a. The entire
network. b. Result of eliminating carnivore 2; the other 10 nodes are unaffected.
c. Result of eliminating the crop; only 6 of the original 11 nodes remain (almost
a 50% reduction).

It is an elementary and common exercise with this sort of representation to
ask the question, how many nodes must be removed before the system collapses?
With such a simple system we can say, as we did earlier, that if the crop and weed
are removed the system clearly collapses entirely. And if we sequentially remove
one node at a time, we can generalize that at any point after the first removal
the probability of collapse with a subsequent removal is about 1%. However,
what seems to interest ecologists more than such calculations is whether the
system is "robust," which is to say, does it resist collapse in the face of removing
nodes? A network consisting of 100 plants and one generalist herbivore would be
extremely robust in the sense that only when the 100 plants have been removed
does the system formally collapse (all species are lost). Contrarily a system of one
plant with 100 herbivores would be very fragile – non-robust – in the sense that
when that one plant is removed, the system collapses completely. This question
is frequently approached empirically for a given network by a sampling scheme
in which a single node is removed and the resulting network examined. This
process is repeated a number of times, until an average node density for a single
removal can be calculated. Then two nodes are randomly removed and the
resulting network examined. Again, the process is repeated a number of times,
until an average node density for a single removal can be calculated. In this way

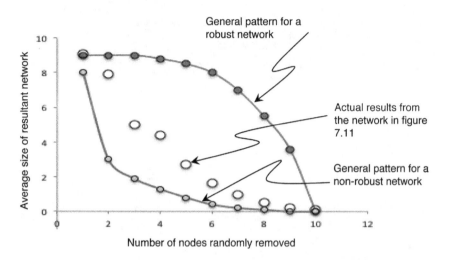

Figure 7.12 Pattern of network structure as a function of number of nodes randomly removed for a robust network, a non-robust network, and the network of Figure 7.11

a graph can be constructed of "number of nodes removed" on the x-axis and the average size of the network on the y-axis. The pattern of change then gives us an idea of the robustness of a system, as illustrated in Figure 7.12.

It is clear that part of the ability to conclude collapse in this artificial example is due to its directed nature and the tacit assumption that there has to be something to eat for each element that needs to eat (e.g., if herbivore 2 in Figure 7.11c were eliminated, both carnivore 2 and parasitoid 3 would disappear also). As we noted earlier, frequently a network is known only to the extent of which nodes are connected to one another, with no suggestion of direction of flow (i.e., a simple graph rather than a directed one). In this case, concluding anything about robustness is far more difficult and requires some other assumption about the definition of collapse or partial collapse.

Qualitative structure of directed graphs – loop analysis

In a sense, much of ecology has been concerned with the very general question of how much do we need to know about a system in order to gain some sense of "understanding" that system. And it is generally assumed that any attempt to manipulate or manage would be greatly aided by such understanding. When complicated, as is normally the case in agroecological systems, the construction of graphs or networks, such as the famous spaghetti graph of the cod system (Figure 1.1), is an attractive approach that at least provides a visual aid. The question, however, is whether making such graphs actually provides any understanding. Simple non-directed graphs, such as those in Figures 7.8 and 7.9 are useful when dealing with, for example, the transmission of pathogens, or migratory patterns among islands or habitats or fields. Yet, as described in the previous section, not much can be said with a non-directed graph when the direction is as important as it is in something like a food web. The use of directed graphs is thus an important step toward the accurate representation of a food web (e.g., Figure 7.11).

When exploring the dynamics of systems represented by directed graphs, there is a tendency to try and ask very detailed questions about the specific behavior of the systems, effectively writing rules (equations) for each of the connections and simply solving those rules on a computer for particular examples. Such an approach is perhaps necessary to eventually make very precise predictions about how a complicated system will unfold over time in particular situations. However, to gain a more intuitive understanding of how structure emerges from a directed graph, there is a variety of fairly sophisticated approaches.[14] One of the most elegant is the system of loop analysis, pioneered by Levins.[15]

To see the basic idea of loop analysis, it is convenient to indicate both positive and negative directions of the connections to the nodes, an example of which is illustrated in Figure 7.13, which is an elaboration of the network in Figure 7.11a. Note that each connector (in network theory lingo, "edge") has either an arrowhead or a small circle or neither, at each end. The arrowhead indicates a positive effect (e.g., herbivore 1 has a positive effect on parasitoid 1, since energy is transferred from the herbivore to the parasitoid). The small circles indicate a negative effect (e.g., all three herbivores have a negative effect on the crop), while no symbol indicates that the effect is trivial (e.g., parasitoid 2 has at most a trivial effect on herbivore 1). This sort of network is a "directed graph" in mathematical terms, and is, we argued earlier, the minimal representation of something as complicated as a food web (recall our original elaboration that an

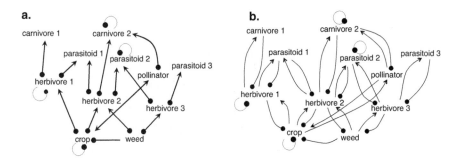

Figure 7.13 a. Directed graph showing the sign of all connections, positive (with arrowheads), and negative (with small circles) and neutral (with no indication). b. The same graph with each direction of interaction (positive or negative) indicated separately.

undirected graph is perfectly appropriate for other applications such as migration or transmission rates among nodes).

Generalizing the picture further, it is easy to see how a general picture such as the one in Figure 7.13 can be generated from some very simple observations about a system, and that is partially the point about the utility of the method of loop analysis. One can see, for example, that carnivore 1 in Figure 7.13a has a positive effect on the crop through its control of herbivore 1. In Figure 7.13b we diagram the same network with both positive and negative effects shown separately, enabling a more detailed examination of the "loops" in the system. So, for example, carnivore 1 has a negative effect on herbivore 1, herbivore 1 has a negative effect on the crop, the crop has a positive effect on herbivore 1, and herbivore 1 has a positive effect on carnivore 1. There is in effect a loop of length 4 from carnivore 1 to itself (operating through herbivore 1).

Analyzing these self-loops is what loop analysis is all about. There are two basic structures relevant to this sort of analysis, the loops themselves and the feedbacks they represent. For example, in Figure 7.14 we have reproduced Figure 7.13b, and indicated various loops to be discussed. Note that herbivore 1 has a positive effect on carnivore 1, which in turn has a negative effect on herbivore 1 – a loop of length 2. Similarly, but with a longer loop structure, the crop has a positive effect on the pollinator, which has a positive effect on carnivore 2, which has a negative effect on herbivore 2, which has a positive effect on parasitoid 2, which has a negative effect on herbivore 3, which has a negative effect on the weed, which has a negative effect on the crop – a loop of length 7 (crop, pollinator, carnivore 2, herbivore 2, parasitoid 2, herbivore 3, weed). Note that the loop of length 2 has a negative sign – a positive effect times a negative effect is negative. We leave it to the reader to verify that the sign of the loop of length 7 is positive.

There is yet one more structure required to understand the basic operation of loop analysis, feedback. Feedback has a colloquial meaning, of course, and that

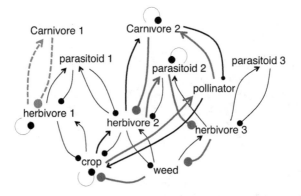

Figure 7.14 Repeat of Figure 7.13b, with a loop of length 2 labeled with dashed connectors and a loop of length 7 with bold connectors

meaning extends to the more formal operation. When the tank in a modern toilet fills, a floating device rises and eventually shuts off the water input, the cycle repeating itself after each flush. The level of water in the tank feeds back to the input valve, telling it to shut off when the water level reaches some critical point. Such feedback mechanisms are so ubiquitous we hardly recognize they are there. But in the context of loop analysis, we must look a bit more carefully at what they mean.

Feedback can exist at a very "low" level, which is to say at the level of a single element (e.g., a population of pests, or a nutrient in the soil). The classic case of a population growing very rapidly at low population densities but then slowing its growth as it gets bigger, eventually approaching no growth as its density approaches the environmental carrying capacity, is a case of feedback at level 1, and by convention we refer to the loop from a species to itself as a loop of length 1 (a single element is involved, feeding back on itself). A predator eating a prey is an example of a loop of length 2 (Figure 7.14).

The system feedback in loops greater than length 1, by definition, includes all feedbacks, which is to say all levels of feedback in the loop. So, for example, the basic process of competition can be visualized as constructed of loops. As we diagram in Figure 7.15a, the overall process (competition) in its classical sense

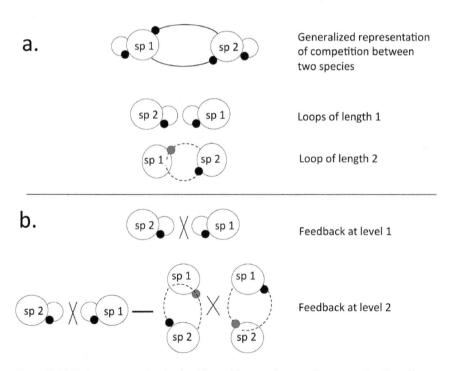

Figure 7.15 Basic structures involved in classical loop analysis. a. Illustrating the idea of loops and their lengths. b. Illustrating the idea of feedback at particular levels (note the horizontal line in the feedback at level 2 is meant to be a minus sign).

includes two loops of length 1 (the density dependent feedback of both species) and two loops of length 2 (the final effect of species one on itself through its effect on species 2 and the final effect of species 2 on itself through its effect on species 1). The feedback is formalized as the sum of the products of all feedbacks. So the feedback at level 1 (staying with the competition example, Figure 7.15) is the product of all the feedbacks at level 1. The feedback at level 2, on the contrary, is the product of feedbacks at level 1 minus the feedbacks at level 2 (Figure 7.15b).

The formal way in which loop analysis proceeds is a bit complicated. One must calculate feedbacks at all levels and decide whether they are positive or negative and then subject them to a set of well-specified rules. The simple example in Figure 7.15 provides a concrete example. We have just two species that are competing with one another and both have self-loops (each has a loop of length 1). The feedback at level 1 is thus the product of the two feedbacks. But there is also feedback at level 2, which is to say species 1, by reducing the growth of species 2, helps itself, so has a positive effect on itself (the loop of length 2, from species 2 back to itself is positive). Exactly the same thing happens for species 2. So, the feedback at level 1 is positive (the product of two negatives is a positive), as is the feedback at level 2 (again multiplying two negatives). But, by the definition of total feedback (see Figure 7.15b), we must subtract the level 2 feedback element from the level one feedback element. Thus, total feedback in the system at level 2 is the difference between feedback at level 1 and the feedback contained in the level 2 loop. If the feedback element for the interspecies competitive effect is less than the feedback at level 1, the system overall is stable. With this example we see how it is necessary only to know the *relative* size of the connections to see if the feedback at loop level 2 is less than feedback at the loop level 1.

Underlying this general qualitative analysis are complicated equations that sometimes allow us to conclude that an overall system is, by its very structural nature, formally stable without knowing exactly the intensity of the interactions but only their existence, direction, and relative sizes. It is an elegant way of looking at a set of interacting species, but is sometimes frustrating in that one of the possible outcomes of the analysis is "don't know." That is, we cannot always conclude one way or another, depending on precisely what the structure of the community is.

Concluding remarks

The acknowledgement that ecological systems are frequently, almost always, composed of many elements (are multi-dimensional) is quotidian.[16] And, as we said initially, it is frequently fodder for the attitude that "there is nothing we can do, there is nothing we can understand." In this chapter we hope we have (1) provided some insights as to the underlying significance of multidimensionality in ecological systems, and (2) provided some potential tools for contextualizing multidimensionality in the context of ecological complexity.

Notes

1 MacArthur and Wilson, 1967; it is worth noting that Levins and Heathwole (1963) had basically noted the idea a few years earlier.
2 Levins, 1969b.
3 This argument is made eloquently for the social sciences by Miller and Page (2009).
4 The late Gary Polis made this point long ago (Polis, 1991).
5 See any elementary population ecology text, e.g., Vandermeer and Goldberg, 2013.
6 Vandermeer, 1992.
7 Frean and Abraham, 2001.
8 Vandermeer, 2015.
9 Vandermeer and Jackson, 2018.
10 Olea and San Miguel-Ayanz, 2006.
11 Originally conceived of by Levins (1968), the community matrix was exploited to great effect in a series of papers by Robert May (1971, 1973, 1974).
12 The idea of a small-world network was originally proposed by Duncan Watts and Steve Strogatz in 1998.
13 A sampling of recent literature includes Newman and Watts, 1999; Want and Chen, 2003; Bassett et al., 2006; Humphries and Gurney, 2008; Sanz-Arigita et al., 2010; Bassett and Bullmore, 2016; Peng et al., 2016.
14 Ranzani and Aumentado, 2015; Chen, 2013.
15 Levins, 1974; Justus, 2005.
16 Examples are not difficult to come by. An excellent example is Rickerl and Francis, 2004b.

8 Trait-mediated indirect interactions

The enemy of my enemy is my friend, and other such aphorisms, reflect an important fact about ecological systems. In 1960 an important paper written by three ecologists from the University of Michigan, professors Hairston, Smith and Slobodkin[1] challenged ecologists to think of terrestrial ecosystems in a highly generalized framework. Known as the "HSS paper," it suggested that some very simple observations could lead to a major conclusion about how ecosystems were structured. Their very simple observations included one that seemed hardly contestable. The terrestrial world looks green. That is, plants, the source of that green color, basically filled up the environment. This observation led them to the conclusion that the things that eat plants, herbivores, were not being limited by those plants (plenty of green stuff around, they argued), and if that be the case, they must be limited by the things that eat them, the carnivores. Their hypotheses became known as "the world is green hypothesis." HSS came to the very strong conclusion that, in general, terrestrial ecosystems had a basic structure, one where the primary producers (plants), were affected by but not limited by the herbivores that eat them, but these herbivores were indeed limited by the carnivores that eat them. That is, terrestrial ecosystems were basically organized along a simple trophic structure of plants to herbivores to carnivores, and that, effectively, the carnivores represented a positive effect on the plants (the enemy of my enemy is my friend), an observation made frequently in the non-scientific literature, for example in a perceptive essay by Aldo Leopold titled "Thinking Like a Mountain," where he described overgrazing by deer after the local extinction of wolves due to hunting.[2]

Evidence of similar phenomena existed also in the aquatic literature from the work of Czech ecologists Hrbáček and colleagues, who demonstrated experimentally that fish in artificial ponds reduce the abundance of zooplankton resulting in a dramatic increase in phytoplankton.[3] The provocation of HSS consisted of taking these simple observations and turning them into a general hypothesis of how communities were structured. Although HSS unleashed a firestorm of criticism, it ultimately led to a whole new paradigm of research, trophic cascades (the effect of a predator "cascades down" through the herbivores to affect the plants, or the resources used by the plants limit the growth of those plants, which limits the herbivores, which in turn limits the predators in a "cascade up" from plant to predator – leading to the frequent use of the terms bottom-up or top-down).

This generalization is actually even more general. Indeed, some of its underlying structure had been anticipated decades earlier by the limnologist Raymond Lindeman's articulation of the idea of trophic levels and the energy dynamics among them (as discussed in Chapters 2 and 7).[4] Indeed, trophic cascades are actually a subset of a larger category of interactions referred to simply as "indirect interactions."

Density-mediated indirect interactions

Lindeman's classic paper on trophic structure not only set the stage for the explosion of an entire subdiscipline of ecology, systems ecology, but contained within its fundamental propositions many structures that have since come to be recognized as keystone properties of food webs, increasingly important as we come to analyze larger and larger such webs. One of those structures was the idea that a consumer has a positive effect on the things its food consumes; for example, a predator is a friend of the grass that its herbivore prey eats. This sort of trophic connection, where a species indirectly affects another through the basic act of consumption, has come to be referred to as a "density-mediated indirect interaction" (to be contrasted to the "trait-mediated indirect interactions" discussed later). A clear indirect mutualism is implied between the plants that the herbivore eats and the predator that eats the herbivore. Although its various modern formulations go by a variety of names, its existence in Lindeman's formulation is obvious. However, Lindeman was not really the first to note these indirect interactions. Darwin's insights in several places in his masterpiece cite this curious fact that effects are not always direct. Even before Darwin, the obviousness of the whole structure was appreciated by ancient Chinese farmers, one of whom pointed out, in approximately AD 300:

> A factor which increases the abundance of a certain bird will indirectly benefit a population of aphids because of the thinning effect which it will have on the coccinellid beetles which eat the aphids but are themselves eaten by the bird.[5]

As noted earlier, modern literature refers to this basic arrangement as a "trophic cascade," which in the end is just one example of what more generally is known as an "indirect effect." The contrasting structure is, of course, a direct effect, in which species A directly affects species B (a predator eats a prey, a parasite attacks a host, etc.). When species A affects species B through its action on species C, the effect of A on B is said to be indirect, perhaps the most obvious and ubiquitous example of which is a simple trophic cascade (predator aids plant by eating herbivores, the plant's enemies).

The question as to whether HSS were right in their claim that the herbivores in terrestrial communities were controlled by their predators, or what has been called "top–down control", as opposed to competition for limited resources, or what has been called "bottom–up control," was hotly debated as a generalization. Furthermore, applying the idea to marine and freshwater communities brought with it some obvious complications. In a major experiment, a lake shaped like an hourglass was divided in two by placing a barrier at the narrow gap in the hourglass, and one side had all the top predator fish removed. The trophic chain here, as in many other freshwater systems, consisted of phytoplankton eaten by zooplankton eaten by small fish eaten by large fish, a clear four-element chain. The very striking result of this experiment was that the side of the lake where the top predators (large fish species) were removed turned green. Obviously, the small fish, released from their predators, increased their populations and ate too many zooplankton, leaving the phytoplankton without control, thus creating a green lake.[6]

Our work in the coffee agroecosystem is illustrative of some of the more complicated results of the many experiments on trophic cascades, yet suggests a potential generalization.[7] To summarize what we discussed in our previous book, *Coffee Agroecology*, we noted what seemed to be a universal pattern in terrestrial communities, at least those communities that do not involve large mammalian predators and herbivores, like lions and zebras. Small vertebrate predators such as birds, lizards and bats ate both invertebrate predators (e.g., spiders) and invertebrate herbivores. So one could imagine the overall structure as one in which predators controlled herbivores, as in the simplified system of HSS, or as one in which top predators (vertebrates) control invertebrate predators (e.g., spiders), which control herbivores that eat the plants, similar to the four-level freshwater system. By constructing exclosure cages to eliminate birds and bats in a coffee agroecosystem, our conclusion was that neither structure was actually operative. Rather, there were significant numbers of parasitoid wasps, too small to be effective food items for birds and bats, but easily captured in the webs of spiders. Based on these results we suggested that the trophic structure consisted effectively of five layers, plants (coffee), eaten by invertebrate herbivores, which in turn were eaten by birds but also attacked by parasitoids, which were eaten by the invertebrate predators (mainly spiders), which were eaten by the birds (Figure 8.1). And, recalling Chapter 6, it is evident that we have here, in principle, a complicated system of six coupled oscillators (birds/spiders, spiders/parasitoids, parasitoids/herbivores, spiders/herbivores, birds/herbivores, herbivores/plants). Given the obvious size constraints (birds or lizards or bats are unlikely to be eaten by spiders, for example), it is a structure that seems likely to occur commonly in all sorts of terrestrial ecosystems. Yet it is a structure that has not yet been investigated theoretically.

The idea of trophic cascades is thus a relatively complicated one when we begin examining it in detail. Yet there is one thing about it that makes it relatively simple, and perhaps unrealistic: the underlying framework of effective linearity. For example, when examining a network such as the one presented in Figure 8.1, it is tempting, if you don't think too deeply about it, to think of the system in a very linear fashion (the plants determine directly how much the herbivores will

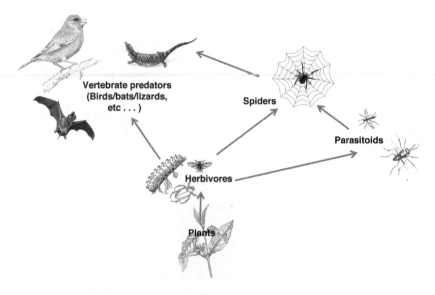

Figure 8.1 The trophic structure deduced from birds/bats exclosure experiments in a Mexican coffee farm, generalized (see Perfecto and Vandermeer, 2015a for details). Such a basic arrangement is likely to be almost universal for terrestrial ecosystems.

increase, the herbivores determine directly how much the spiders will increase, and so on). As much as we understand it to be an oversimplification, it is tempting to assume that the per capita growth in the population of herbivores is simply directly proportional to the biomass of the plants, at least to a first approximation. Yet we see, even from a superficial examination of the network, that the growth of the herbivore population is also dependent on the predators of the herbivores. And then the "enemy of my enemy is my friend" framework, frequently referred to as an "indirect interaction," further complicates even this simple linear generalization. But there is something far more complicated (or complex) that is usually at play in natural systems, including, of course, agroecosystems. The indirect interactions may involve more than just a concatenation of direct interactions. They may involve the modification of traits other than the simple transfer of energy (which is basically what the arrows in Figure 8.1 refer to). Those indirect interactions that involve transfer of energy are usually referred to as "density-mediated indirect interactions". Those that involve the modification of traits are referred to as "trait-mediated indirect interactions" (TMII).[8]

Trait-mediated indirect interactions in principle: a basic nonlinearity

A story from eighteenth-century North America provides an almost trivial example of what trait-mediated indirect interactions are. In rural North

America, before massive industrialization, it was a commonplace situation that young boys were excused from classes for the purpose of "scaring crows." In the fall, when grains were filling in the massive monocultures of the Midwest, crows were a continual menace. For some period of time the standard way of dealing with this pest was for the young boys of the family to go out and scare the crows away (Figure 8.2). An artificial scarecrow would help, but boys with slingshots or noisemakers were clearly more effective. And it is obvious that, although they may have killed a crow or two with a well-placed slingshot, the effectiveness of their operation was the degree to which they "scared" the

Figure 8.2 The Bird Scarer, by William Knight Keeling (1807–1886), painted between 1827 and 1886. Depicting a boy holding a noisemaker in his hand to scare away the crows that were a constant pest in eighteenth and nineteenth-century farming.

Source: Wolverhampton Art Gallery, reprinted with permission.

crows – the degree to which they could change the crow's behavioral trait of eating. Their impact on the crows was mediated through this change in behavior.

If there had been thousands of boys with shotguns in a given field, we might imagine that it would be the actual killing of the birds that had the effect. It would still be an "indirect" effect in that the predator (the human boys) were indirectly affecting the maize by killing the birds, and the more boys with shotguns the more birds they would kill. So the indirect effect could be thought of as mediated by the number (density) of boys with shotguns. Mediating the indirect effect of human on maize through a change in the behavior of the pest contrasts with the density-mediated indirect effect and is thus referred to as a trait-mediated (in this particular case the trait is the behavior of the birds) indirect effect.

A well-known issue in the intercropping literature is, in the end, a perfect example of the distinction between trait-mediated and density-mediated indirect effects. Historically small-scale producers have routinely combined crops in the same fields,[9] and more recently the growth of modern agroecological applications has rediscovered (sometimes reinvented) this traditional management style.[10] Among the many reasons for the practice, the potential for control of pests is frequently cited. In a seminal paper in the 1970s, Root and colleagues[11] elaborated two alternative hypotheses regarding pest control in polycultures, (1) the resource concentration hypothesis and (2) the natural enemies hypothesis. These clear alternative explanations for reduced pest problems in intercropped fields generated considerable interest among experimentalists.[12] On the one hand, the "resource concentration hypothesis" suggests that herbivores, some of which are potential pests, become less able to locate their host plant (the crop) because of some disruption caused by the presence of an alternative crop (e.g., cabbage moths cannot find their host plants as efficiently if interspersed tomato plants produce volatiles that overwhelm the cues produced by the cabbages). On the other hand, the "natural enemies hypothesis" suggests that the alternative crop(s) provide resources (e.g., flowers and nectar) that either attract predators from outside of the system or provide alternative energy for those predators to increase in population density. The predators then consume the potential pests (e.g., parasites of cabbage moths utilize the flowers of the tomatoes and thus increase their local densities, thus potentially controlling the pests).

There is, in this early research program, a clear example of the distinction between density-mediated and trait-mediated indirect interactions. It is evident that the enemies hypothesis is an example of a density-mediated indirect interaction since the alternative crop causes an increase in local density of natural enemies that then consume the herbivores. On the other hand, the resource concentration hypothesis is an example of a trait-mediated effect, which is to say, the alternative crop reduces the oviposition rate of the potential pest, as depicted in Figure 8.3.[13]

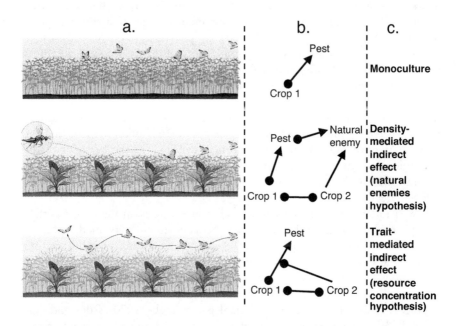

Figure 8.3 Schematic of the effect of a second crop (the darker plant) on an herbivore of the primary crop (the lighter plants). a. Pictorial representation of the three alternatives wherein a lepidopteran pest attacks a monoculture (top diagram), is itself attacked by a predator or parasite whose population is increased by the presence of the secondary crop (middle diagram), or is repelled or confused by signals from the secondary crop (bottom diagram). b. Interaction diagrams illustrating the direct and indirect interactions involved with the three alternatives. Arrowheads indicate positive effects, closed circles indicate negative effects. Note in the bottom graph the impact of Crop 2 is on the interaction term describing the relation between Crop 1 and the pest. c. Statement of each of the three alternatives both as the indirect effects and the original pest control hypotheses (in parentheses). Drawing by John Megahan.

A slightly different lesson is provided by the well-known mutualism between ants and various hemipteran herbivores, one of the most commonly cited mutualisms in nature. The hemipterans produce a sweet substance, called honeydew, that the ants consume. The ants protect the hemipterans by scaring away (or sometimes eating) their potential predators and parasitoids. So, the basic energy transfers are hemipteran to ant, through the honeydew, and hemipteran to predator or parasite. If we view energy transfer as similar to water flowing through tubes, in this example we can illustrate the water flowing (1) from the plant to the hemipteran, (2) from the hemipteran to the ant and (3) from the hemipteran to the predator, as in Figure 8.4a. With this simple arrangement we presume that the per capita production of offspring by the hemipteran is a fraction of the energy

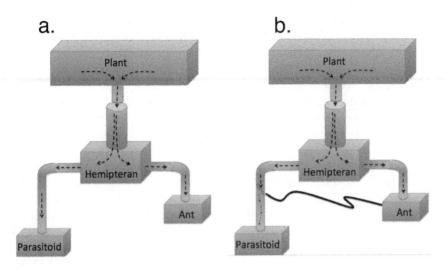

Figure 8.4 Flow diagram illustrating the transfer of energy and the effect of trait-mediation in a simple system. a. Direct uninhibited flow from the plant to the hemipteran herbivore to the ant and to the parasitoid. The ant takes energy in the form of the honeydew secreted by the hemipteran and the parasitoid takes energy in the form of the hemipteran biomass as it is being consumed. b. Same as (a), except the ant enables a "valve" in the tube that stop the flow of energy from the hemipteran to the parasitoid. The effect of the valve is the trait-mediated effect, and adds an essential non-linearity to the system.

it obtains from the plant minus a fraction of the energy it gives up to the ant and the parasitoid. So, we could write the equation,

per capita change in hemipteran $= a(\text{plant}) - b(\text{ant}) - c(\text{parasitoid})$

where a, b and c are proportionality constants, and "plant," "ant" and "parasitoid" represent the biomass or density of each of those elements. This is a linear arrangement. Now if we add the effect of the ant, we see that the presence of the ant actually changes the rate (c), which is to say the rate at which the parasitoid takes energy from the hemipteran and the basic equation then [after adding $c = c'(\text{ant})$] becomes,

per capita change in hemipteran $= a(\text{plant}) - b(\text{ant}) - c'(\text{ant})(\text{parasitoid})$

whence we see that the ant biomass is multiplied by the parasitoid biomass. This adds an essential "non-linearity" to the system (Figure 8.4b).

Note that the indirect interaction of the ant with the parasitoid in Figure 8.4b is very different from the indirect interaction of the birds with the plant in Figure 8.1. The latter is an indirect effect mediated through the direct energy channels, a density-mediated indirect interaction (the impact of birds in general will increase as the density of the bird population increases), while the former is an interaction mediated by a change in some trait (in this case the feeding efficiency of the parasitoid), and thus it is a trait-mediated indirect interaction. And the most important fact of trait-mediation is that it usually implies an essential non-linearity added to the system.

And we must emphasize, this is not a small discardable idiosyncrasy that occurs perforce, providing interesting material for popular articles and television shows. There are many examples in which the trait-mediated indirect effects are far more important than the direct effects, one example of which is the common association of ants with hemipterans (Figure 8.4). As noted in an especially enlightening article[14] considering potential interactions, as the dimensionality of the system increases, the trait-mediated effects increase at a far greater rate than the direct and density-mediated effects. For example, in a three-element system, groups A, B and C make three direct interactions (AB, BC, AC) and three potential trait-mediated interactions (A affects the interaction BC, B affects the interaction AC, and C affects the interaction AB). With four groups there are six direct interactions (AB, BC, CD, AD, AC, BD) and 12 potential trait-mediated effects (A affects the interactions of BC, CD and BD, and the other three, B, C, and D, also each potentially have three interactions to influence). With five groups there are six potential direct effects and 30 potential trait-mediated effects. Thus we go from a 1:1 ratio with three groups to a 2:1 ratio with four groups to a 3:1 ratio with five groups. And the deviation keeps building: 4:1 for six groups, 5:1 for seven and so forth. But if we then add the potential for "cascading" effects, (e.g., A affects BC, but D affects the ability of A to affect BC) the potential for these trait-mediated non-linearities becomes enormous – the lion scares away the gazelle so the cheetah can't catch it, but the humans scare away the lions, interfering with their ability to scare away the gazelles, but the elephants attract the tourists, relieving the pressure on the lions. Darwin's famous observations reflect this ever-spiraling cascade:

> bumble-bees alone visit the common red clover (*Trifolium pratense*), as other bees cannot reach the nectar. Hence I have very little doubt, that if the whole genus of bumble-bees became extinct or very rare in England, the heartsease and red clover would become very rare, or wholly disappear. The number of bumble-bees in any district depends in a great degree on the number of field-mice, which destroy their combs and nests; . . . Now the number of mice is largely dependent, as everyone knows, on the number of cats; and [it is said] "Near villages and small towns I have found the nests of bumble-bees more numerous than elsewhere, which I attribute to the number of cats that destroy the mice." Hence it is quite credible that the presence of a feline animal in large numbers in a district might determine, through the intervention first of mice and then of bees, the frequency of certain flowers in that district![15]

Huxley, with tongue in cheek to be sure, added:

> old maids [unmarried elderly women] keep cats, and . . . the economy of the British Empire [is] based on roast beef eaten by its soldiers, and cattle rely on clover, so [we must] conclude that the prosperity of the British Empire was thus dependent on its population of old maids.[16]

Of course both density-mediated effects (the cats eat the mice) and trait-mediated effects (the mice destroy the bee nests) are involved in this fanciful connection of international politics with pollination.

Putting such burlesque aside, there are actually two issues meriting consideration. First, what is the extent of trait-mediated effects in the real world, and second, what are the consequences of these effects on the structure and function of the ecological systems they drive?

Trait-mediated effects in the real world

A colleague of ours, Earl Werner, has been emphasizing the importance of trait-mediated effects since before they seem to have been officially baptized (our research on this may be historically incomplete, but the phrase itself seems to have been originally promoted as "interaction modifications" by Wootton in 1993, although the identical idea of "higher order" effects is well-known in other literature[17]). The classic example[18] involves two size classes of bullfrog tadpoles (*Rana catesbeiana*) in competition with one another, one of which (the smaller class) is attacked by a dragonfly predator (*Anax* sp.). If there is a large number of smaller tadpoles in the system, their collective algae grazing leaves little for the larger tadpoles to eat, and the latter will die – that is, unless the key predator also lives in the pond. Smaller tadpoles have much to fear from the dragonfly predators, so, in the presence of the predator they reduce their foraging activity (since they have to attempt to hide from the predator), thus releasing the larger tadpoles from the competitive pressure. So far, this story is identical to many three-species systems in which a predator preferentially consumes one resource that is in competition with a second resource, thus indirectly benefiting that second resource. But the system is more complicated in a very interesting way. Through a series of clever experiments, it was found that the odor of the predator initiates a behavioral response in the smaller tadpoles such that they hunker down in the mud, dramatically reducing their rate of consumption of the algae, thus altering their competitive effect on the larger tadpoles. So the predator need not eat the tadpole to change the competitive effect. Its very presence, which the tadpole senses chemically, is enough to do so. This is a classical trait-mediated indirect effect.

An example from our own work in the coffee agroecosystem illustrates how these trait-mediated effects can become quite complicated, and reinforces the notion that they may indeed be more important than the direct effects. The centerpiece of this example is the same ant species we already introduced in Chapter 3,

Azteca sericeasur. The ant feeds on the honeydew secreted by the scale insect, *Coccus viridis*, a potential coffee pest. In a typical ant-hemipteran mutualistic interaction, the ant scares away (and sometimes kills) the beetle, *Azya orbigera*, which would normally kill/eat the scale insects. The larval stage of the beetle, however, has a waxy cuticle that the ant is unable to penetrate. The larva is thus immune to the attacks of the ant and consumes the scale insects in great quantities. And remember, beetle adults generally fly, so the adults can fly around looking for scale insects to eat, but the larvae are confined to a local area and unable to move around very much. Consequently, the larvae need to be located in a local concentration of scale insects to survive. This presents a problem for the beetles, since the adults must place their eggs in the vicinity of the scale insects, indeed in a place where there is a local concentration of scale insects. But the only places where there are concentrations of scale insects are places where the scales are tended by the ants. This represents two problems for the female beetle that is trying to oviposit. First, the ants will not allow the adult beetle anywhere near the scale insects and second, if the beetle does manage to lay an egg close to the local concentration of scales, the ants will immediately remove it.

The second problem is solved ingeniously by the female beetle. She lays her eggs underneath the scale insects, where they are safe from the attacking ants. This species of scale insect engages in a sort of parental care in that the offspring (formally known as "crawlers") are maintained under the shell of the mother, thus gaining protection from their potential natural enemies. Except, of course, for the predaceous beetles that have managed to deposit their own eggs underneath the body of an unsuspecting scale. But the question still emerges regarding the first problem: how to get an egg, and resultant larvae, within a local concentration of scale insects in the first place, where the ants are constantly patrolling and harassing any organism that gets close to the scales? The answer to this question involves a remarkable complication that includes a new trait-mediated indirect interaction.

The ant is attacked by a parasitic fly in the family Phoridae. The phorid parasitoid is able to detect the general locality of ants by using the alarm pheromone of the ants. But in order to be able to successfully oviposit an egg on the ant, the phorid needs to sense movement on the part of an individual ant. Without movement, the phorid is effectively blind. Therefore, as a defense strategy in response to this parasitoid, the ant has evolved a peculiar behavioral response; when the phorid arrives, the ant assumes a catatonic (immotile) posture (unless it can rapidly escape into its nest), such that the phorid cannot "see" it. But the ants have also evolved a sophisticated communication system with one another with regard to this parasitoid. The first ant that is attacked by a phorid emits a special pheromone that basically says "look out, a phorid has been sighted close by," and all the other ants in the vicinity assume that same catatonic posture. Therefore, when the phorids arrive in a particular area where the ants are tending scales, the activity of the ants is reduced dramatically. It is precisely at that point that the beetle is able to penetrate the defense the ant offers the scale insect, move to the concentrations of scale insects, and deposit eggs underneath the scales.

But how is the gravid female beetle to know that window of opportunity has arisen? If she keeps trying to penetrate the active ant defenses, waiting for that moment when all the ants are catatonic, she keeps exposing herself to attack by the ants. It turns out that the beetles have evolved the ability to detect the particular ant pheromone that instructs the ants to get paralyzed because the phorids are nearby. In other words, the beetles, able to decipher the communication among the ants, are eavesdropping on them and respond when an opportunity arises. What is fascinating is that female beetles, but not males, respond to the "look out, a phorid has been sighted close by" pheromone. But not only that, only females that have already mated do this, which is to say non-gravid females purposefully ignore it or are even repelled by it. Furthermore, the more days after copulation, the more sensitive is the gravid female to this special pheromone. [19]

The system is obviously quite complicated and, we argue, it is a perfect example of a complex system. We summarize some of its complexity on a point-by-point basis as follows:

1 The ant, *Azteca sericeasur*, consumes the honeydew produced by the scale insect, *Coccus viridis*.
2 While consuming the honeydew, the ant scares away or directly attacks any adult individual of the beetle, *Azya orbigera*, that tries to get near the scale insects. Thus the hunting behavior trait of the beetle is modified by the activity of the ant.
3 Larval beetles have protective waxy filaments making them impervious to the attack of the ants. Thus the trait of the ants to attack anything that comes near to the scale insects is modified.
4 Larval beetles, due to their relative immobility, must be located within a heavy concentration of the scale insects to survive. Adult beetles, in contrast, are able to fly and forage on isolated scale insects not yet protected by ants.
5 The ants clean the leaves of the coffee plants, including any beetle eggs that may be in the vicinity, making the problem of oviposition for the beetle serious. Oviposition needs to happen near heavy concentrations of scale insects, a situation that never arises unless under the protection of ants. Thus, the very local density of the scale insects modifies the oviposition trait of the beetle.
6 The beetle has solved the problem of oviposition by ovipositing underneath the scale insects, but must face the problem of ants attacking ovipositing females. Thus the trait of cleaning leaves (which would include cleaning them of beetle eggs) is modified by the behavior of the beetle.
7 A phorid fly, *Pseudacteon* spp., attacks the ant. But the ant has a defensive behavioral response that renders it catatonic for a period of time (from 0.5 to 1.0 hours). The foraging trait of the ant is modified by the attacking phorids.
8 The ant's defensive catatonic response results from a special pheromone produced by the ant at the time a phorid is detected in the area, causing

all ants in a local area to assume the catatonic position, thus providing a window of opportunity for female beetles to lay eggs. Thus the behavior of the ant in producing this special pheromone modifies the trait of oviposition of the phorid parasitoid.

9 In order to recognize the window of opportunity, gravid female beetles have evolved the ability to detect the special pheromone the ant produces to signal the presence of the phorid. Effectively the beetle eavesdrops on the chemical messaging of the ants. Thus, the beetle's oviposition behavior is modified by the production of this special pheromone.

More simply, using the basic idea of energy flowing through pipes but modified by controlled valves, we illustrate a few of the basic elements of this system in Figure 8.5. It is obvious, we think, that the energy flow in the system (the food web of classical ecology) is far less important to the operation of the system than the cascading trait-mediated effects. That is, the dashed arrows, while a representation of reality, are not nearly as important as the "wires" controlling the "valves" and "switches" in the system.

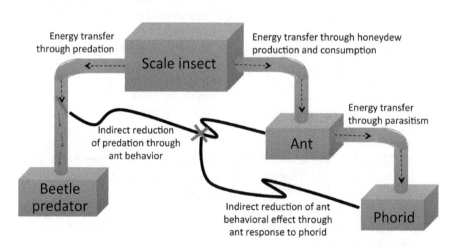

Figure 8.5 Diagrammatic representation of the ant-scale-beetle-phorid trait-mediated trophic cascade. Metaphorical tubes carry energy and metaphorical electric lines operate valves and switches to change the flow of energy. Note that the underlying biology of the system indicates that "energy transfer through honeydew production and consumption" and "energy transfer through parasitism" are completely trivial compared to "energy transfer through predation" in this figure. Furthermore, the trait-mediated indirect effects of the valve closing and switching (the black "wires" in the diagram) completely overwhelm any of the direct effects in this example. Another aspect of the interaction, not included in this simple representation, is the oviposition behavior of the beetle, in which gravid female beetles are able to detect the signaling of the ants that a phorid is in the area.

Consequences of trait-mediated indirect interactions

There is a major, if not fully appreciated, consequence of the existence of TMIIs in nature. To fully appreciate this consequence, it is necessary to quickly review one of the major results of community ecology, previously alluded to briefly (Chapter 7). First analyzed by Levins and more or less simultaneously expanded by May,[20] at the time it was a remarkably surprising result. There had been an underlying assumption that ecologists had made, largely from a background of natural history and nature writers. The ecosystem was stable, so the argument went, because of its complexity and a major measure of that complexity is the species diversity. Thus, biodiversity per se became associated with the stability of the ecosystem. A metaphor that was frequently used was that of a spider web gaining its strength from its many connections. Even today many popular outlets tacitly assume that the stability of an ecosystem is directly related to its diversity.

This idea was challenged by Levins and May. They noted simply that if you take the classic theoretical formulations of community ecology, and do the calculations, you find that stability DECREASES as biodiversity increases. This result was based on some simplifying mathematical assumptions, but despite attempts at rescuing the old idea of biodiversity begets stability, there was no way of avoiding the conclusion that the reverse was the theoretical truth. Ecological communities were not like spider webs. A more useful metaphor would seem to be a house of cards: the more cards, the more unstable the structure becomes.[21]

The existence of TMIIs, as we indicated, challenges this result and, we believe, reopens the question of how biodiversity relates to stability. Two general approaches have recently been brought to the table: hypernetworks (an extension of basic network theory) and higher-order interactions (the idea that a trait-mediated effect is effectively an extra non-linearity in the system). We discuss these two in turn.

Hypernetworks

In the previous chapter we introduced the concept of networks (or graphs of systems) as a way of approaching systems that are inherently multidimensional. The presence of trait-mediated indirect interactions presents a basic challenge to this approach. Consider, for example, the simple relationships of the *Azteca* ant system presented in Figure 8.5. A classical network approach might begin with some very simple assumptions about connections, presuming four "nodes" and three edges (Figure 8.6a). This is the approach taken in much of the literature on network analysis, usually dealing with very large networks such as electricity grids or the internet, where the connection itself is what is important, not its direction or intensity. One might ask the question, how much can we know about a complicated system if we understand nothing more than which node is connected to which other node? This interesting question dominates the literature on very large-scale networks. However, when dealing with smaller networks it is sometimes the case that knowledge of what is "connected to

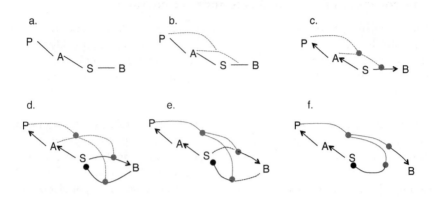

Figure 8.6 Various network representations of the *Azteca* system that include the phorid (P), *Azteca* ant (A), scale insect (S), and beetle (B). a. A simple non-directed graph of the system, perhaps the simplest network that might be visualized – the amount of information contained here is limited, to say the least. b. Addition of the trait-mediated interactions to form a "hypergraph." c. Addition of directionality to the hypergraph; the arrows indicate the flow of energy (direct interactions) while the small circles indicate "negative effect." d. Incorporating all population effects; solid connections indicate energy transfer (toward the arrowhead, away from the circle), dashed curves indicate trait-mediated effects (circles at end indicate negative effect). Similarly, the (e) elimination of the sections of the direct effects that are "intercepted" by the trait-mediated effects. f. Including only the important effects in the system, whereby it is easy to visualize the dynamics determined mainly by the indirect effects (the phorid reducing the ant's ability to interfere with the beetle). Note that the final effect is easily visualized by following each path. From phorid to beetle we encounter two circles, indicating two negative effects, plus one arrowhead, indicating a positive effect, making an overall positive effect of P on B. From phorid to scale we have three circles, indicating three negative effects and an overall negative effect of P on S. Note that the phorid has a negative effect on itself through the pathway P-S-A. Graphs involving such constructed nodes are referred to as bipartite graphs.

what" does not provide enough information to understand the system. Adding the trait-mediated effect to the elementary graph produces a "hypergraph"[22] (Figure 8.6b). A qualitative look at the hypergraph suggests that the simple graph could very well be missing key elements, elements that may very well overwhelm the dynamics suggested by the simple graph, as is the case in the *Azteca* system described earlier (Figure 8.5). More information can be provided by adding directionality to the hypergraph, such that the direction of energy flow and control (the trait-mediation) are evident (Figure 8.6c). Indicating the direction of all interactions adds yet further information to the network (Figure 8.6d and 8.6e), which finally may result in a "stripped-down" graph that suggests where the main dynamic influences of the system are (Figure 8.6f). Clearly the phorid is having a major effect on the system, and that effect is largely due to the trait-mediated effects. Knowing only about the connections or even the directed connections, reveals little about what is ultimately driving the system.

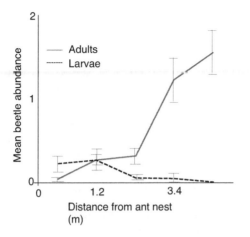

Figure 8.7 Density of adults and larvae of the scale-eating beetle, *Azya orbigera*, as a function of distance to nearest nest of the ant *Azteca sericeasur*

Source: From Liere et al., 2014.

Recalling the basic natural history of this system, it is a system that exists in a spatial context that is a key aspect of the system dynamics (Chapter 3). In the case of our ongoing studies in the coffee agroecosystems in Mexico, the ants nest in shade trees within the coffee farms and exist in clusters or aggrega- tions of trees that cover between 3% and 10% of the potential space. Careful surveys reveal that the beetle adults are virtually absent within about a meter of the tree containing the ant nest, but increase in abundance dramatically up to almost 5 meters. Beetle larvae, contrarily, have their highest abundance within 1.2 meters of the ant nest (Figure 8.7). The adult beetles are very efficient at locating individual scale insects through olfactory cues, and range widely on the farm, and undoubtedly beyond. Larvae, as we noted earlier, are largely restricted to a few branches of an individual coffee plant (where the concentration of scales are), which means they need those high concentrations of scale insects to survive, which only occur when the ants are there. So, in the end, even though the scale insects may reduce the production of coffee on the particular bushes where they are feeding, they simultaneously create "sources" of adult beetles that can range widely and effect almost complete control on scale insects on the rest of the farm. The scales thus provide a base for their own control, and it all happens because of a combination of direct and indirect effects, including a cascade of trait-mediated indirect effects. In Figure 8.8 we made an attempt to incorporate all of this natural history into the hypergraph approach.

Turning now to the dynamic consequences of TMIIs, we first note some important specific results. As in the case of the ant–hemipteran mutualism, some classical ecological interactions have been misinterpreted as being direct, even

a. b.

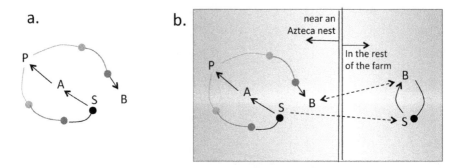

Figure 8.8 Representation of the control of the scale insect pest in the coffee agroecosystem.
a. Hypergraph representation, the identical topology to Figure 8.6f. Note that P
and B are connected with a hyper edge and P and S are likewise represented with
a hyper edge. b. Representation of the system in space, where the patch "near an
Azteca nest" provides a source for the adult beetles that range widely across the
farm and effectively control the scale insect pest. Dashed lines represent dispersal
events.

though their basic structure suggests the dominance of a trait-mediated effect.
And we cannot emphasize enough the fact that an extra nonlinear term is always
added when the trait-mediated effect is in force (this is a subclass of what had
in the past been called higher-order effects). While the general consequences of
adding these trait-mediated effects is still in the exploratory stage from a theory
point of view, we do know that the effects have been reported regularly in the
literature. Golubski and Abrams,[23] for example, discussed a series of categories of
TMIIs that have been reported in the literature. They note that one of the most
frequently reported TMII is the modification of a predator prey relationship in
the context of a trophic chain. Consider the familiar example in Figure 8.9. The
cheetah is an efficient hunter of gazelles, but when the gazelles have tall grass
to hide in, the hunting efficiency of cheetahs is reduced. Likewise, when there
are cheetahs in the vicinity, the gazelles, forever nervous about being attacked
by cheetahs, reduce their feeding rates. This kind of dual effect on the basic
predator/prey relationship is common.

An alternative scenario might be the operation of two agents that impose
distinct TMIIs on a system, and sometimes those TMIIs may be antagonistic to
one another as suggested in Figure 8.10. The tall grass provides a hiding area for
the gazelle, thus decreasing the rate at which the cheetah can prey upon it. But
there is a disease that attacks the gazelle, making sick individuals easy prey for the
cheetah, thus increasing the rate of energy transfer from the gazelle population
to the cheetah population (Figure 8.10).

In their review of the literature on TMIIs in nature, Golubski and Abrams
conclude that first, the aggregate effect of distinct modifier species (e.g., the tall
grass and disease organism of Figure 8.10) differs from the independent combination of single modifier effects. This means that different TMIIs are sometimes
interdependent on one another. Second, there is a meaningful structure to the

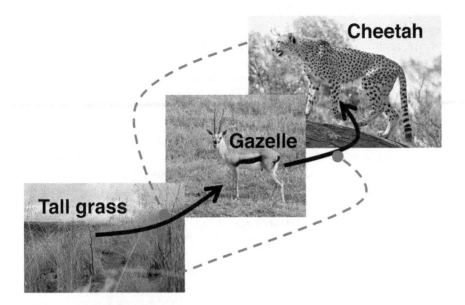

Figure 8.9 Illustration of the two classes of TMII in a generalized trophic situation. Black arrows indicate direct energy transfer; dashed curves indicate negative changes to the rates of energy transfer. Note that there are two generalized classes of TMIIs included here: modification of the rate of transfer of energy to the herbivore by the way the predator changes its eating habits, and modification of the rate of transfer of energy to the predator by the way the prey is able to hide in the tall grass.

Figure 8.10 TMIIs with alternative effects on the transfer of energy from gazelle to cheetah

multiple TMIIs. For example, modifier species that act to modify a particular energy transfer modality (e.g., spiders trapping insects in webs) are likely to impose similar TMII effects on all the prey items they trap in their webs. Third, and perhaps a bit speculative, they propose that "interactions between TMIIs usually cause each modifer's effects . . . to be weaker than would be expected if TMIIs were independent" (Golubski and Abrams, 2011, p. 1106).

From this third conclusion one can speculate that the overall effect of TMIIs will be most important for ecosystems of intermediate size (which is the case for most agroecosystems). Very large systems will see the effects of TMIIs attenuate since many of them may interact to cancel one another, but very small systems may not be large enough to effectively "sample" the potentially available TMIIs in the species pool. The effects of this insight on agroecosystems could be important. In the industrialized agricultural model, the obsessive search for simplification (the holy grail of which is the industrial grain monoculture) suggests that, at least in the abstract TMIIs may not be very important. At the other extreme, however, traditional agroecological farms typically have what seems like a very large biodiversity, both planned and associated. Nevertheless, at least in tropical systems, the overall biodiversity of associated natural systems is almost always much greater than the agroecological systems, suggesting that perhaps a characteristic of them is precisely the intermediate scale in which it is suggested that TMIIs are most important.

Multiple TMIIs

Although the general consequences of adding trait-mediated effects is still in the exploratory stage from a theory point of view, a recent theoretical analysis of TMIIs provides us with a seemingly major result, in some ways challenging the results of Levins/May concerning the relationship between biodiversity and stability.[24] Categorizing interactions as either two-way, three-way, or four-way (the latter two of which are what we refer to as TMIIs), the question was asked, at some particular size of the community (number of species interacting) what is the critical size of the interaction coefficient that leads to instability (in this case, from computer runs, sometimes one or more species is eliminated from the community). Below that critical size all species (at that particular size of the community) remain in the system in perpetuity, but above that critical size, some species go extinct. The basic setup is illustrated in Figure 8.11. Constructing random systems with either only two-way interactions or with only three-way interactions or with only four-way interactions (see Figure 8.11), we can ask what is the pattern of critical interaction strength as a function of the number of species in the system.

Extensive simulations using randomly chosen parameters (i.e., choosing α, β and γ randomly), it seems to be the case that with only two-way interactions, the classic result is, not surprisingly, repeated (more species leads to greater instability). However, adding the three-way and, especially, the four-way interactions resulted in the opposite effect – as the number of species increases, it is more likely that the system will be stable. In Figure 8.11 we illustrate, with a particularly simple example, how this result seems to emerge.

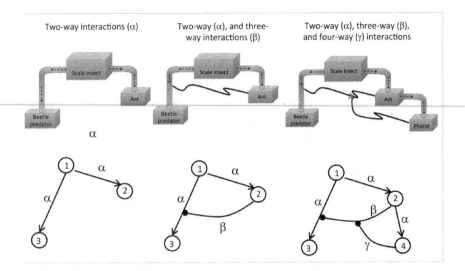

Figure 8.11 The framing of the higher-order effects using the framework from Figures 8.4 and 8.5. a. The ant-scale-beetle-phorid system from the coffee agroecosystem in Mexico illustrated as a flow diagram. b. Formulating the trait-mediated indirect interactions as multi-way higher-order interactions (α = two-way interactions, β = three-way interactions, γ = four-way interactions), where the species numbers corresponding to the actual species in part a are 1 = scale insect, 2 = ant, 3 = beetle predator, 4 = phorid fly.

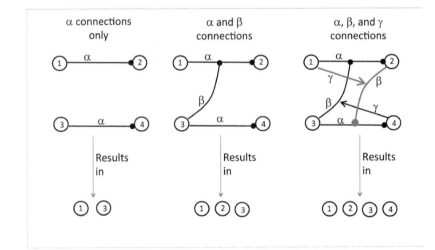

Figure 8.12 Example of the general outcome of simulations with α, β and γ connections. With only simple two-way connections (α connections), species 1 and 3 dominate in competition with species 2 and 4, and the result is a system composed of only species 1 and 3. With three-way competition (α and β connections), with the particular arrangement shown here, the strong competitive effect of species 1 on species 2 is partially curtailed, resulting in species 1, 2 and 3 surviving in the final system. In four-way competition (α, β and γ connections), with the particular arrangement shown here, the strong competitive effects (the α effects), are curtailed by the β effects, which are reinforced by the γ effects, resulting in all four species being retained in the system.

Summary

Indirect effects have long been appreciated in ecology (the friend of my enemy is also my enemy, etc.), with the basic trophic chain of carnivore-herbivore-plant being a classical example, forming a foundation of much research in community ecology (e.g., trophic cascades, bottom-up versus top-down effects). More recent literature has compounded these insights, noting that traits other than population density can have significant effects on the structure and dynamics of communities.[25] Thus we have seen the emergence of a research program contrasting density-mediated indirect effects versus trait-mediated indirect effects. The latter are now thought to be of great importance in structuring nature, so much so that they are sometimes thought to be more important than the direct effects. Several research paradigms, both empirical and theoretical, have resulted in a rich literature emphasizing the potential importance of such effects. However, what's important for us, from the perspective of this book, is that these higher-order (trait-mediated) indirect effects add non-linearities to the system, making them complex systems. Such trait-mediated indirect interactions are common in natural as well as agricultural systems.

Notes

1 Hairston, N. G., Smith, F. E., and Slobodkin, L. B. (1960). Community structure, population control, and competition. *American Naturalist*, 421–425.
2 Leopold, 1949.
3 Hrbáček et al., 1961.
4 Lindeman, 1942.
5 Needham, 1956, cited in Boler, Van Lenteren and Delucchi, 2006. Pg. 10. (http://docplayer.net/21772588-Iobc-internet-book-of-biological-control-version-6.html) accessed 23 December 2016.
6 Schindler et al., 1996.
7 This is certainly a complicated issue, further discussed in Chapter 7. The important article of Polis (1991) both clarifies and criticizes our approach.
8 Werner and Peacor, 2003; Abrams, 2007; Golubski and Abrams, 2011.
9 Altieri, 1999; Francis, 1986; Gliessman, 1987, 1990; Vandermeer, 1989.
10 Vandermeer, 1992, 1995; Power, 1989; Ramert et al., 2002; Altieri, 2004; García-Barrios, 2003.
11 Root, 1973; Root and Kareiva, 1984; Tahvanainen and Root, 1972.
12 Bach, 1980; Risch, 1981; Kareiva, 1982; Risch et al., 1983; Russell, 1989.
13 What is frequently not noted in this example is that the additional crop, whether acting in a resource concentration fashion or an enemy attractive fashion, does incur a cost to the principle crop through potential competition. This aspect has not been fully treated in the literature.
14 Golubski and Abrams, 2011.
15 Darwin, 1859, chap. 3.
16 Huxley, 1892.
17 Any system that uses differential equations as dynamic models recognizes this concept. See, for example, Vandermeer, 1969.
18 Peacor and Werner, 2001.
19 Hsieh et al., 2012; Liere and Larsen, 2010.
20 Levins, 1968; May, 1973b.

21 For those unfamiliar with this metaphor, as one balances playing cards on top of one another at various angles, it is possible to create large structures, houses, that grow more unstable the bigger they get, eventually toppling from their own weight.
22 Golubski et al., 2016.
23 Golubski and Abrams 2011.
24 Bairey et al., 2016.
25 Many examples are presented in the various chapters in the edited volume by Ohgushi et al., 2012.

9 Critical transitions

Inevitability of surprise

As people come to internalize the assertion that ecosystems are almost infinitely complicated, a sense of foreboding enters their expectations of the future, not sure what surprises await them, and certain that the ecological world will not remain constant. Farmers have long been familiar with such uncertainty. Their legitimate worries and persistent insecurities stem from the evident fact that farming has a certain level of unpredictability, and for this reason, it is a risky business. For the most part, the risks are thought to be mainly due to inherent unpredictabilities in the economic system, such as price fluctuations and the unpredictability of the weather. But these uncertainties take on new meaning when ecological complexity is added to the equation. That is, when we realize, for example, that the millions of bacterial cells in the soil interact with the mycorrhizal fungi, which determine the internal chemistry of the crops which affects the rate of feeding of pests, which themselves may or may not be strongly attacked by parasitoids, which sometimes get caught in spider webs made by spiders, that sometimes get eaten by birds, which need fruiting trees all year to survive, and on and on. Concatenating contingencies can seem overwhelming. The sensitive dependence on initial conditions that the chaos revolution alerted us to, or the complex coupled oscillators that are common to all ecosystems, loom large as we face the realities of the complicated connections among thousands of elements in an average agroecosystem. Therefore, it should not be surprising when we claim that unanticipated behavior, surprises, are likely to be inevitable in all ecosystems, including the agroecosystem. But rather than taking on the attitude of *lo que será, será*, we ask whether it is possible to find structure in the very idea of surprise.

There is, fortunately, an evolving framework for understanding at least some of the structure of surprise. That framework originally stems from an obscure collection of abstract mathematical theorizing referred to as catastrophe theory.[1] We can get an idea of how the framework works with an examination of the classical problem of biodiversity reduction as a function of agricultural intensification.

Loss of biodiversity: expected or surprise?

There is a seemingly impenetrable bastion of thought among many conservationists about the relationship between agriculture and biodiversity. It is easy to
see where this bastion originates. If one looks at a GMO soybean field, or the
massive expansions of industrial banana or palm oil plantations in the tropics,
or the "amber waves of grain" spreading to the horizon in North America, it
is simply a no-brainer: agriculture "kills" biodiversity. Yet when forced to think
more deeply about what agriculture actually is, this framework becomes a bit
altered. Take, for example, the production activities of the Huaorani people in
the Amazon rainforest.[2] They are mainly hunters and gatherers and thus have a
practical stake in conserving natural biodiversity, yet they too engage in a form
of agriculture. Their settlements are temporary, but last long enough to leave a
signal in the forest. One piece of that signal is the several individuals of peach
palms they leave among their temporary dwellings, trees that germinate from
the seeds they discard after eating the fruit and grow relatively rapidly, but not
fast enough to produce harvestable fruit before the family moves on to another
site. Thus, to harvest their crop, they must return to the site of previous temporary settlements, which may turn out to be far removed from where they live.
Indeed, many trees that were planted, perhaps one or two generations ago, must
be located within the matrix of rainforest and harvested, almost as if they were
just another species of fruit in the forest. Is this agriculture? By definition it must
be, but no one would suggest that such a modification of the habitat actually
results in significant loss of natural biodiversity (indeed, at a very local level, it
may actually increase the biodiversity of some plants and animals).

   Huaorani agriculture is only an extreme example. The *dehesa* system, as we
noted in the previous chapter, contains a remarkable amount of biodiversity,
probably quite similar to the original vegetation of the area, which was undoubtedly oak savannah. Agroforestry systems, such as traditional cacao or coffee, are
known to be important repositories of local biodiversity, not just the biodiversity planned by the farmer, but the biodiversity that eventually associates itself
with that farming operation.[3] The idea that it is the kind of agriculture, not
just its existence, that translates into different amounts of biodiversity is now
well-known,[4] and can be easily visualized using the concept of an "intensification gradient." We can, for example, imagine an agroforestry system that ranges
from very dense tree cover, approximately equal to nearby forests, to a less dense
canopy of overstory trees, to only a few scattered trees of the same species, to,
in the extreme, no trees at all, just the crop (think of going from the Huaorani
peach palms, to the *dehesa* system, to the shaded coffee system, to a citrus grove,
to a maize monoculture). This sort of gradient, from something similar to native
vegetation at one extreme, to something similar to an open wheat field at the
other extreme is what we refer to as an intensification gradient.[5] It is evident that
the trend to be expected is that biodiversity will generally decline as this kind
of agricultural intensification proceeds.

A largely discredited early literature in conservation biology tacitly assumed an extreme: that as soon as agriculture appears on the scene, biodiversity would drop precipitously. The truth is that the implied dramatic loss of biodiversity with the imposition of agriculture is a bit more nuanced than the average conservationists might think at first glance. The pattern may be as suggested in Figure 9.1a. Almost as soon as agriculture is introduced, biodiversity declines very rapidly. The reality is more nuanced, and there are some systems and some groups of species for which the pattern shown in Figure 9.1b, is most common.

It is tempting to think of the two patterns illustrated in Figure 9.1 as a complete catalogue of how agricultural intensification could be related to biodiversity. However, there is clearly another pattern, a bit more complicated but clearly possible. For example, imagine a farmer replacing an industrial monoculture with a diversified agroecological farming operation. To achieve that, the farmer will stop applying agrochemicals, plant fruit trees, timber trees, and nitrogen fixing trees. Later, she will intercrop a rich diversity of vegetables with cover crops, and will leave some of the natural vegetation to regrow on the edges to attract natural enemies. Eventually, we may expect that birds will find the trees and nest in them, and beetles will encounter the fallen logs that inevitably result as the trees age, and small mammals will eventually find refuge in the deep leaf litter that the overstory trees produce. However, it will take some time for all

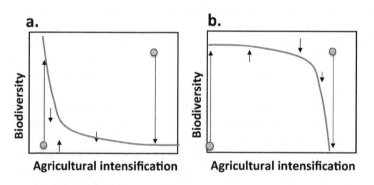

Figure 9.1 Biodiversity changes in response to agricultural intensification. Framing the idea as a deterministic relationship between agricultural intensification and biodiversity, wherein each point on the x-axis (agricultural intensification) produces the corresponding biodiversity on the function that relates the independent variable (agricultural intensification) to the dependent variable (biodiversity). Note that the arrows (vectors) in each of the panels indicate the expected change from the base of the arrow to its point. Especially note the arrows with small circles at their beginning. a. The frequently presumed relationship between biodiversity and agricultural intensification, in which only something as benign as planting fruit trees in a forest allows biodiversity to be conserved in the face of agriculture. b. Alternate framing in which there is a smooth transition from high to low biodiversity as agriculture becomes intensified, and the major losses in biodiversity occur at higher levels of intensification.

of that to happen, and at least the initial result of starting that new farm will be some increase, but not much, in biodiversity. In contrast, imagine a similar farming operation but initiated with the modification of an extant forest. In that case, the farmer will cut some trees and plant others, and will probably leave some of the natural vegetation on the edges of the farm. Later he will open up enough canopy to plant some vegetables, and engage in more intensive tillage. In this case, the birds are already there, as are the small mammals and beetles. The initial biodiversity will be substantially larger than in the first example – the same farming operation starting on land that had been previously in industrial monoculture cannot be expected to result in the same biodiversity as when started by modifying the natural vegetation. As the intensification proceeds (as the farmer continues adopting modern methods of chemical usage, tillage, crop planning, etc.), slowly but surely the birds that flew from tree to tree will disappear, the beetles will find fewer rotting logs to burrow into, the small mammals will find the more open understory less desirable. Biodiversity will decrease, perhaps slowly, but decrease it will.

In the first case we have very little biodiversity at first, but as a more ecological system is established, the biodiversity begins to increase, while in the second case, as a more intensified industrial system is approached, the biodiversity begins to decline. We can imagine a point where both farms have pretty much the same agricultural intensity, but the first will have lower biodiversity because of its legacy while the second will have higher biodiversity also because of its legacy. Thus we must conclude that a single point on the intensification axis can result in very different amounts of biodiversity.

A useful way of visualizing this phenomenon is by taking pieces of the alternative framings (i.e., those in Figure 9.1), and suppose one represents intensification beginning from the natural system (top panel in Figure 9.2a) while the other represents a "de-intensification" from the intensive industrial system (bottom panel in Figure 9.2a). Note how these two segments of the relationship overlap, which is to say there are some values of intensification (some values on the x-axis) for which there are alternative equilibrium points. This means that if you start the system with low levels of intensification and high levels of biodiversity and start intensifying the system (moving to the right), the high levels of biodiversity will be maintained for a while until you reach a "critical" (tipping) point of intensification (on the right-hand side of the graph), where biodiversity suddenly collapses to very low levels. Alternatively, if you start the system with high levels of intensification and low levels of biodiversity and start de-intensifying the system (moving to the left), it will remain at low levels of biodiversity until you reach another "critical" (tipping) point of intensification (on the left-hand side of the graph), where biodiversity suddenly increases to high levels. However, if you start the system in between those two critical points, the biodiversity will either be high or low, depending on where it actually starts when the farm is set up (where it is for the particular value on the x-axis, as indicated by the long arrows beginning with the lightly shaded circles in all the panels of Figure 9.2).

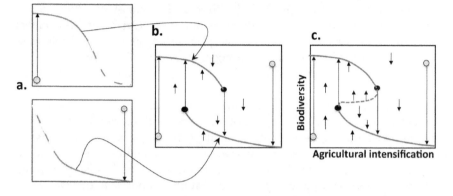

Figure 9.2 The construction of tipping points and hysteresis. a. Repeat of the two curves in Figure 9.1, indicating how the construction of the panel in (b) is a combination of the two functions in (a) (also Figure 9.1b). b. The construction of the new function using the partial elements of the original two functions (note long narrow arrows connecting line segments in (a) with the segments of the function in (b)). Note the inevitable emergence of two critical points (arrows with black ovals at their start). c. Final construction of the "fold catastrophe," in which the dashed curve indicates the locus of unstable points in the system, where the darkly shaded circles indicate points of critical transition. Note that at each point on the *x*-axis within the region between the two critical points is a region for which two alternative stable equilibrium points exist, separated by an unstable point (all the unstable points are indicated by the dashed curve and all the stable points indicated by the solid curves). This region is referred to as the hysteretic region. The extreme points (represented by lightly shaded circles) are the same in (b) and (c) (as well as in both panels of Figure 9.1), indicating that in all this development the extremes are the same, with very small amount of agricultural intensification there is large biodiversity while with large amount of intensification, there will be a small amount of biodiversity, no matter where you start the system.

A glance at Figure 9.2b reveals a logical difficulty within the region where there are two alternate states. If biodiversity begins relatively low, it will descend to the lower leg of the function. If biodiversity begins relatively high, it will ascend to the upper leg of the function (note the small vectors indicating the underlying dynamics). There must be some point where biodiversity is more or less balanced between the tendency to ascend and the tendency to descend. This is a classical unstable equilibrium point (see Chapter 2). For every point on the *x*-axis between the two critical points, there must be an unstable equilibrium point located between the two stable points. The position of those unstable points is indicated with a dashed red curve in Figure 9.2c.

This framing provides us with a convenient metaphor for surprise, sudden unexpected change. In particular it suggests how biodiversity might change, indeed, how it might change suddenly and surprisingly as agricultural intensification proceeds. How precisely that pattern emerges from basic principles is not so straightforward, which is to say that the mechanism producing the pattern shown in Figure 9.2c is not completely transparent. In other applications underlying mechanisms may be even less obvious, even though tipping points are suspected to occur somewhere within the dynamics of the system. Furthermore, we may suspect that an ecological system contains important tipping points, but it may not be evident precisely where those tipping points occur, and this is what makes the change a surprise. In any particular case the existence and even approximate location of tipping points, and thus our ability to predict them, may be deducible from an understanding of the underlying dynamics of the system of concern. An example of how the pattern of critical transition can emerge from first principles is provided by a classical ecological idea of vegetation response to precipitation.

Hysteresis on a global scale

Consider the tree cover in tropical landscapes, one important determinant of which we know is precipitation. As we move from low to high precipitation, the ecosystem tends to have a greater density of trees, and it moves from savannah to forest, as described many times in elementary ecology texts. However, in the past, this view had been accompanied by a tacit assumption – that the function translating the condition (moisture) to the response variable (density of trees) is single-valued. That is, for every value on the x-axis there is only one value on the y-axis – as in Figure 9.3a and 9.3b). The relationship between the condition and the response variable could be more or less linear, like the example in Figure 9.3a, indicating that a certain change in the condition will generate a proportional change in the response variable. But it can also be non-linear, like the example in Figure 9.3b, indicating that a small change in the condition can generate a disproportionally large change in the response variable, in other words, a critical transition. More recent literature acknowledges that many ecological trans-formations are not at all like this, but rather something more like illustrated in Figure 9.5c, where there is a zone in which alternative states are possible for a particular value of the condition, with a broad zone of hysteresis. Indeed, many ecological examples have been proffered illustrating this pattern (e.g., turbidity in lakes,[6] tropical savannah/forest conversion,[7] marine phytoplankton).[8] In ecology, different states generally are referred to as "regimes".

Returning to the example of tree cover, we see that if the annual rainfall begins to change, as we might expect in the near future due to climate change, if the relationship is as pictured in Figure 9.3a, as annual rainfall declines, we expect a sort of concomitant reduction in tree cover, progressing more or less in lockstep with changes in precipitation. However, the evidence we have so far is that such a smooth transition with a particular change in rainfall leading to a similarly proportional change in tree cover, is not likely to be the reality. More likely is a situation as pictured in Figure 9.3b, in which there is a zone of rapid transition, where a small change in rainfall leads to a large change in tree cover. In this case, a surprise is likely to happen at some point, where previously small responses in tree cover to a particular change in rainfall suddenly give way to a large change in

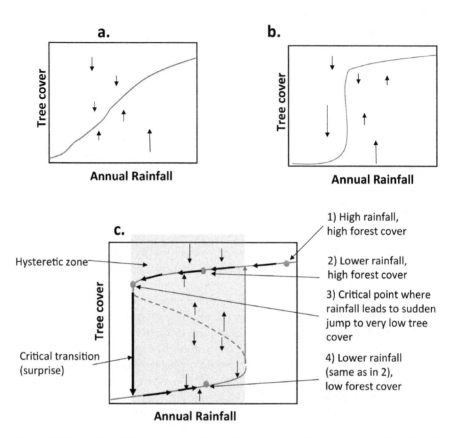

Figure 9.3 Examples of the qualitative relationship between ecological regime and ecological conditions. a. What is sometimes assumed without much reflection, that changes in conditions, in this case, annual rainfall, produce more or less proportional changes in a response variable, in this case, tree cover. The "regimes" range continuously from high tree cover (forest) to low tree cover (savannah). b. Adding nonlinearities to the system suggests that some changing conditions will induce more rapid changes in the response variable than other changing conditions. c. Relaxing the assumption of a one-to-one relationship between the conditions and the response variable, produces a range of conditions for which there are two stable "regimes" (forest and savannah), with a broad hysteretic zone on the annual rainfall axis.

tree cover. This represents to some degree an underlying structure for surprise, or a critical transition. However, a more interesting structure is illustrated in Figure 9.3c, where a hysteretic zone is present.

The form illustrated in Figure 9.3c (the same as was qualitatively derived in the case of the biodiversity in agriculture in Figure 9.2c) is classically known as a critical transition graph with hysteresis (as the previous example, Figure 9.2). Imagine that you are located in a tropical area with a large annual rainfall, such that you would expect very high tree cover (see the point labeled "High rainfall, high forest cover" in Figure 9.3c). Due to global climate change we now see a reduction in annual rainfall in the region such that on the graph we move according to the three arrows pointing left on the function, until we arrive at the point of "lower rainfall, high forest cover." Note that not much changes on the *y*-axis as we move from high precipitation to low precipitation since we remain in the high tree cover region of the *y*-axis. However, further decreases in rainfall (illustrated by the fourth arrow pointing left) quickly reaches a "critical point". At this point there is still a stable forest situation, but even a very slight reduction in rainfall results in a very sudden (and perhaps surprising) reduction to very low tree cover, a savannah situation. This is the structure that, in our view, is a metaphor for many sudden and surprising changes in ecosystems. And note, we humans are not likely to know the precise underlying function and thus may be completely incapable of anticipating these surprises.[9]

Analyzing the critical transition graph further we find an even more surprising structure. Having arrived at the point of very low rainfall and savannah (after the critical transition has occurred) if now the rainfall begins increasing again, even though we may reach a point of rainfall that previously had given rise to a forested environment (high tree cover), we now are "stuck" with a savannah (see point number four in Figure 9.3c). There is a zone for which alternative equilibrium states are possible for any of the values on the *x*-axis. This zone is referred to as a "hysteretic zone," and strong empirical evidence exists for such a zone in tropical terrestrial communities.[10]

Yet a broader reflection suggests a further complication. When thinking of precipitation and the responses of natural vegetation, thoughts frequently go beyond the forest to savannah transition. Indeed, a coarse appreciation of tropical vegetation includes not two but three general categories, forest, savannah and desert. The same arguments about precipitation apply when thinking of the savannah/desert transition, such that Figure 9.3c could be just as easily applied there, only at a smaller scale of precipitation. Thus, considering the whole range of possible precipitation regimes, we might expect something like Figure 9.4a. And examining the actual data for much of the tropics we see a broad correspondence with the basic idea.

Thus we see at a global level there is considerable support for the idea that distinct regimes of tropical vegetation exist, and, further, that those regimes frequently exist in a hysteretic state. And the dynamic consequences are disturbing to many. As is made clear in Figure 9.4, if precipitation drops below the tipping point, many areas of tropical forest will suddenly turn into savannahs and many areas of tropical savannah will suddenly turn into deserts with further reduction in precipitation. But the most disturbing aspect is not so much the expected critical transition, but rather the hysteresis. We could, for example, imagine that humanity suddenly comes to grips with the problem of climate change and engages in a massive effort to reduce CO_2 (transform energy systems to renewables, change agricultural systems, plant tree, etc.), such that the precipitation regime changes to what it had been before the critical transition. This could have

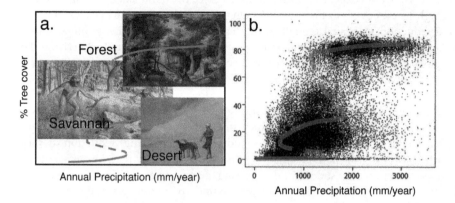

Figure 9.4 Relationship between annual precipitation and percent tree cover creates distinct ecological regimes. a. The theoretical relationship (note that the probability of fire is a component in the real world, but not included here to keep things simple – annual precipitation has to be sufficiently large to create enough vegetation for kindling a fire, but if too large, the vegetation remains too moist for fires to spread.). Original art added to base images: a. Jan Brueghel, *Forest Landscape with Deer Hunt* (image at https://commons.wikimedia.org/wiki/File:Jan_Brueghel_%28I%29_-_Forest_ landscape_with_deer_hunt.jpg); Australopithicenes (https://commons.wikimedia .org/wiki/File:Australopithecus_family.jpg); Jean-Léon Gérome, *On the Desert* (https://en.wikipedia.org/wiki/Jean-L%C3%A9on_G%C3%A9r%C3%B4me#/ media/File:Jean-L%C3%A9on_G%C3%A9r%C3%B4me_-_On_the_Desert_-_ Walters_3734.jpg). b. Collating annual precipitation and tree cover information over 1 km² grid cells between 35°S and 15°N in Africa, Australia, and South America.[11]

Source: Hirota, M., Holmgren, M., Van Nes, E. H., & Scheffer, M. (2011). Global resilience of tropical forest and savanna to critical transitions. *Science, 334*(6053), 232–235.

no effect at all on reversing the critical transition. We may become "stuck" in the ecological regime we unknowingly imposed by short-sighted political decisions.

While the underlying mechanisms of water, fire and vegetation for feeding fires is qualitatively obvious in the forest to savannah to desert system, the details are not necessarily all that well-known, which is to say the detailed mechanisms that produce the fire, the droughts, the proper rain, the microhabitats for seed germination, and a host of other factors that are all involved in the process, are not fully incorporated into our understanding. A smaller scale system, the coffee rust disease, provides us with an example of how ecological regimes form complicated tipping points and hysteretic patterns, from a complicated set of ecological forces.

The surprise of the coffee rust disease

As we began writing this book, a disaster hit the agroecosystem we are most familiar with. The coffee ecosystem of Latin America was hit with a devastating outbreak of the infamous coffee rust disease. Enormous press coverage ensued globally, warning everyone that the world's most popular drug could soon be in short supply. The disease had already made itself famous at the end of the nineteenth century when the British had to give up coffee production in Ceylon altogether after the disease completely devastated their crop – that is why you may have heard of Ceylon tea, but not Ceylon coffee, since tea replaced coffee as the major export of the British colony of Ceylon (today Sri Lanka).

A bit of recent history is useful to set the stage for this example.[12] After abandoning coffee production entirely in Ceylon (and much of southern India, Sumatra and Java), a burgeoning of coffee production took place in Latin America. From Mexico to Brazil, and throughout much of the Caribbean, Latin America became a major source for the world's coffee. In most areas production systems were based on the dominance of the small family farmer, but deeply dependent on cheap harvest labor. Most importantly, with the exception of Brazil, the style of production imitated the natural system where coffee originated (mid-elevation tropical forest), and thus it was usually grown under a canopy of shade trees. And to be sure, everyone was cognizant of the fact that the rust disease, completely unknown in the Americas, remained a looming threat as a worldwide transportation system enabled diseases of all sorts to become world travelers.

From the late nineteenth century until 1966, a cautious optimism reigned throughout the region. The rust remained absent. Then the rust arrived in Brazil, perhaps carried in the atmosphere from an outbreak in Angola, and rapidly gained the attention of both farmers and extension agents. A small army of phytopathologists searched for a solution to the problem, with not much success. Finally, in the early 1980s it arrived in Central America, bringing with it panic and paranoia. The paranoia was perhaps justified given the well-known devastating history of the disease in South Asia. But then the history becomes interesting largely because the problem sort of disappeared. After initial "hair-on-fire" attention, it appeared that the disease settled down to be constantly present, but rarely a major problem. Farmers simply ignored it as it appeared regularly at the beginning of the wet season, increased mildly throughout the wet season, and then disappeared in the dry season. The fear and paranoia seemed to be completely unwarranted.

Then, without apparent warning, in 2012 from Mexico to Peru, the coffee rust emerged as an epidemic, with losses of production reported as high as 40%, a complete "surprise" to most analysts. Given that coffee is the mainstay of hundreds of thousands of families in the region and the economic backbone of several nations, intense attention was riveted on the problem. What caused the sudden epidemic? A new strain of the rust? Maybe. Climate change? Maybe. Coffee losing its resistance? Perhaps. But also it could have been a natural progression of small environmental changes that slowly built up and reached a tipping point. It is that possibility that we shall explore.

A basic understanding of the biology of the coffee rust disease is a necessary prerequisite for seeing its tipping point behavior. In Figure 9.5 is a diagrammatic representation of the life cycle of the disease. Note that the process of dispersion,

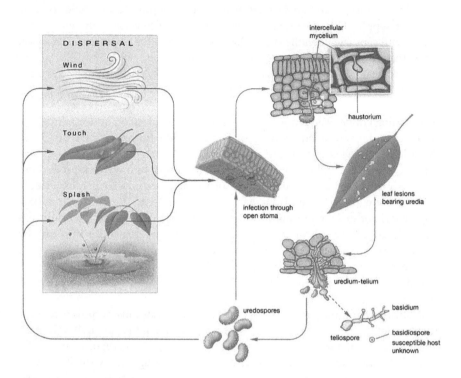

Figure 9.5 Basic life cycle of the coffee rust disease
Source: Diagram made by John Megahan.

that is to say the transmission of the spores of the fungus, is key to the spread of the disease, with three main modes of dispersion involved: wind, contact and splash. Most important for our purposes is the multiple spatial scales of dispersal. Wind carries spores very long distances. So from the point of view of an individual farm, spores are blown in from very far away and we can think of them as a "rain" of spores. But in addition to wind dispersal, spores are transferred from plant to plant by contact on a given farm. This means that in farms where the coffee bushes are planted very densely, the disease can spread rapidly from plant to plant within the farm. In addition to wind dispersal and leaf-to-leaf contact, the spores can be transmitted by splash from the soil. However, this third form of transmission probably contributes more to reinfection of a particular plant than to plant-to-plant transmission. Thus there are effectively two distinct scales of transmission, one large scale transmission manifested as a "rain" of spores over a very large region, and one locally from plant to plant.

Once in contact with the coffee leaf, the spore needs a minimum amount of moisture to germinate (effectively, a given spore requires a drop of water). The local habitat conditions, especially the general conditions of moisture are important variables in determining whether the spore will germinate or not. And once it germinates it penetrates the leaf through the stomata and grows intercellularly, forming haustoria, which are the organs through which the fungus sucks out the nutrients from the spores.

When ready to form new spores, the fungus creates uredia that penetrate through either the same or other stomata. The uredia (or uredium-telium) house the uredospores (or simply spores), which then get blown around by the wind, or rub off onto other leaves if there is contact from leaf to leaf. At this point there are at least two major predators that attack the spores as they are being formed. First, there is another fungus, the white halo fungus (*Lecanicillium lecanii*), that eats the spores and second there is a very abundant fly larva (*Mycodiplosis hemileiae*) that consumes vast quantities of spores. So, in this system, there are three modes of dispersal, a set of habitat conditions that promote spore germination and at least two potential biological control agents.

In Latin America the style of coffee production varies from farm to farm. In its most traditional form, a coffee farm has large numbers of shade trees forming a canopy above the coffee bushes, while the more "modern" farms are coffee monocultures with no trees (Figure 9.6).[13] A dense overstory canopy effectively acts as a windbreak, making it difficult for the rain of spores from the coffee rust to penetrate very far into the understory where the coffee bushes are (Figure 9.6a). Additionally, in this style of production the density of coffee plants is relatively sparse, resulting in minimal dispersion of spores from plant to plant. So, the windbreak effect of the shade trees coupled with the sparseness of the coffee bushes themselves create a condition where the disease itself cannot prosper well.

Figure 9.6 Two extremes of coffee production systems. It is easy to imagine what the wind speed conditions, and thus the dispersal of rust spores, might be in the two examples. a. Typical shade coffee production in southern Chiapas, Mexico. b. Typical sun coffee production in the south of Minas Gerais, Brazil.

Source: Authors.

At the other extreme of the spectrum of coffee production systems is coffee monoculture, or what some refer to as "sun coffee," with few if any shade trees and very high densities of coffee plants (Figure 9.6b). In Latin America, starting in the 1980s, there has been a tendency to change from the shaded system to the open sun system. This has generated a change at the landscape level within coffee landscapes. It does not take much imagination to speculate as to the likely consequences of this transformation for the transmission of the coffee rust. No shade trees means that the wind can disperse whatever spores are in the atmosphere, and the dense planting means that bush-to-bush transmission can be very high. Because of these factors operating together, the theoretical expectation is that the disease will come to be epidemic as the shade trees are eliminated.

The style of management also may have an effect on the potential biological control agents, in particular a mycoparasite, the white halo fungus, that attacks both insects and fungi. In particular, *Azteca* ants (recall the pattern formation in Chapter 3) are mutualistic with scale insects (as described in Chapter 8), so concentrations of scale insects occur only when they are under protection of the ants. The white halo fungus attacks the scale insects, but only when they are locally concentrated. The very same white halo fungus that attacks the scale insects also attacks the coffee rust fungus. The local buildup of scale insects when under protection of the ants forms a concentrated source for the white halo fungus and thus a source area from which to disperse and attack the rust.[14] And, since the *Azteca* ants nest in shade trees, any reduction in shade tree density implies a change in the density of the ant nests and concomitant reduction of this biological control agent.

The interesting part of this story occurs in an intermediate situation, either as we move from a shaded system toward a sun system or from a sun system toward a shade system. If a traditional farm with a high density of overstory shade trees begins the process of converting to sun coffee, the spores begin to move into the farm from the atmosphere at a greater rate than when the shade created a windbreak. So the expectation is that the incidence of the disease will increase, but not necessarily right away or every season. It depends on the state of the farm with regard to the disease already in place, at the beginning of the wet season (the season when the disease is active). This implies that there is some point at which the disease could be epidemic on some farms but very low on other farms, depending on where it began. And, logically, there must be a special point between the two extremes where the potential for these two alternatives emerges. This point is a critical transition point for the disease.

Beginning at the other extreme, a pure sun farm, there is maximum dispersal of spores both from the overall region and locally from bush to bush. As the farm begins the process of adding shade trees and reducing the concentration of coffee bushes in the understory, it moves toward the shaded situation. And, as before, there is some critical point (not the same point as when we moved in the other direction), for which the disease could be epidemic or not, depending on its state at the beginning of the wet season.

The general dynamics of the disease, considering only two factors, the regional transmission rate and the local transmission rate, can be envisioned with a series of drawings as in Figure 9.7, where we see how the change in percent cover of

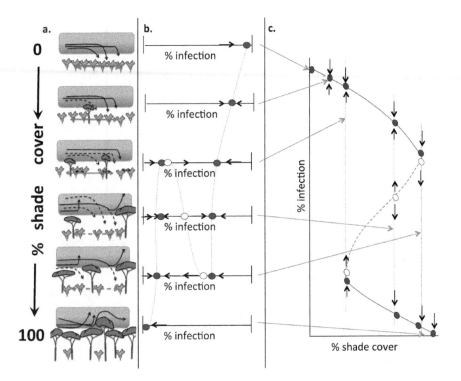

Figure 9.7 Illustration of the changes in expectation of the coffee rust disease, assuming that the transmission of spores is the only factor affecting the disease. a. A pictorial representation of moving the management system from pure sun (topmost picture) to 100% shade. The rounded rectangles represent the atmosphere and solid black and dashed arrows represent spores being blown by wind, either strong transmission with solid lines or weak transmission with dashed lines. Horizontal double arrowhead lines at coffee layer indicate transmission from coffee bush to coffee bush, with weight of line and dashed line indicating relative strength of transmission. b. Expected percentage infestation of the disease on a farm based on the corresponding illustrations in a. Note the two types of equilibrium points, stable or unstable (unfilled). Arrows indicate dynamical changes expected, with both stable and unstable equilibrium points visible. The light line indicates what might be the expectations for intermediate arrangements, connecting homologous stable and unstable points. c. Critical transition graph of the system, where each of the cases in (b) is indicated. Note that the general function is basically a rotation of the connecting lines in (b), where solid line indicates stable equilibria and dashed line indicates unstable equilibria.

shade trees (from top to bottom in Figure 9.7a) is translated into the percent infection of the disease (in Figure 9.7b). The result (Figure 9.7c) is the classical result of critical transitions with a hysteretic zone (the same as described above for the intensification gradient and biodiversity and terrestrial vegetation; see Figures 9.2 and 9.4), and more generally explains the sort of underlying dynamics that can lead to surprises.[15]

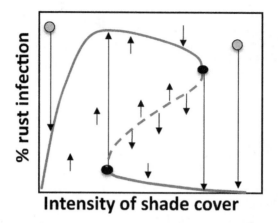

Figure 9.8 The expected relationship between coffee rust disease and the amount of shade
cover in a coffee agroecosystem, based on an underlying model that includes two
spatial scales of transmission, shade-induced transmission and germination effects,
and potential role of natural enemies

However, when taking the other aspects of the disease into account (germination conditions, natural enemies), the story becomes more complicated.[16] Shade,
for example, has two effects: (1) reduction of wind and thus spore transmission,
but also (2) increasing the humidity of the local environment, thus encouraging
spore germination. Furthermore, at least one of the natural biological control
elements is encouraged by the shade (the white halo fungus). If we add these
elements to the system, the picture in Figure 9.8 emerges.[17]

The idea of surprise has evident practical implications. However, the related
idea of hysteresis may ultimately have even more practical importance. In the case
of the coffee rust disease, for example, if the shade cover is one of the determining
factors (many analysts feel this is the case, but it remains controversial), we see
how removing shade trees from the system could lead to an epidemic of the coffee rust disease. However, returning to a system of partial shade may not have the
expected effect on controlling the disease right away because the system has been
transferred to a different "regime" in which the disease is relatively endemic. That
is, we have jumped from the lower branch of equilibrium points in Figure 9.8
to the upper branch. Increasing the shade tree density may not have much of an
effect at all since the system is likely to stay in the epidemic zone (upper branch
of equilibrium points in Figure 9.8), until we reach the new critical point.

Agricultural syndromes as hysteretic phenomena with tipping points

With respect to agroecosystems more generally, distinct regimes, frequently
referred to as "syndromes of production,"[18] are clearly notable at several scales.
For example, the regime called "coffee farm" exists throughout the montane

tropics, interspersed with farms in the regime called "cattle pastures," yet the coffee systems themselves are highly varied, and can be divided into the regimes of "sun" coffee and "shade" coffee, as well as intermediate stages (Figure 9.6).[19] What makes a farmer decide on a particular regime versus another? How easy is it to change from one regime to another? These are important questions for farmers' livelihoods as well as for the environment. We can cast this "regime change" question as a potential critical transition problem in search of understanding how some agroecological regimes (syndromes) are transformed. The basic idea is to take a measure of the regime (for example, coffee at one end, pasture at the other; shade coffee at one end, sun coffee at the other; complex agroforestry system at one end, monoculture of rice at the other) on the y-axis and socio-political conditions that influence farmer's decisions on the x-axis. This is basically the intensification gradient we used in the analysis of biodiversity regimes in the first section of this chapter. But here we are viewing the same gradient, the intensification gradient, not as an input variable to produce something else (e.g., biodiversity as in the previous example), but rather as a response to other conditions, to a different variable, a variable related to farmers' decision-making process.

Here we are considering what is a rather vague notion of the input variable (the independent variable), the set of sociopolitical conditions that result in cacao versus pastures, or shade coffee versus sun coffee, or agroforestry versus grain monoculture and so forth. Farmers must use some set of rules to decide on management practices. From a typical modern economic perspective those rules usually involve calculations associated with current and expected gross and net income, with various forms of scaling on income, and other economic considerations. This economic perspective on the agrarian economy, dominant today but at least visible perhaps since the nineteenth century, was criticized by the Russian economist Chayanov (1966), who intensively studied the Russian peasantry previous to and at the time of the 1917 Revolution, when only a small minority of the Russian population was urban. Chayanov's insights have resurfaced in recent years[20] with the suggestion that rural populations the world over seem to operate in ways that seem qualitatively similar to the general principles elaborated by Chayanov.

The Chayanovian system, elaborated by Van der Ploeg for contemporary farming systems, is based on a complex set of "balances." Two of the most important balances are the "labor/consumption" balance and the "utility/ drudgery" balance. At one extreme of the labor/consumption balance is the situation when farmers and their families work to produce only what they consume – no market involved. The utility/drudgery balance is the intuitive notion, implicit in the thoughts of almost all farmers, that increase in marginal utility declines with increasing production, but simultaneously, the drudgery of achieving that production increases. There must be some point at which the farmer decides that the next unit of utility is not worth the next unit of drudgery, suggesting a local equilibrium. Other balances include farm/nature, local/regional, peasant/ruler, and others. The point of the Chayanovian system is that all of these balances enter into the complex decision-making process of the peasant system.[21]

Historical changes in agriculture qualitatively follow as if they were a function of the conditions, herein taken to be approximated by the degree of deviation from Chayanovian balances, under which management decisions are made. For example, a crude, very broad brush of European farming history suggests a transition from (1) a long time frame of pure peasant farming, to (2) expanding markets with exchange value intermingling with use value, to (3) political pressure to move further toward exchange value, suggesting increasing area farmed, to (4) development of extensive monoculture to take advantage of new innovations in mechanization, to (5) concentration of land power resulting from attempts at more efficiency, to (6) further chemicalization of pest control and soil fertility, to (7) the need for more land to justify investments in inputs, to (8) elimination or marginalization of the peasantry. Such an historical trajectory, or something similar, has been frequently repeated all over the globe to one extent or another. Formally, there is no reason, other than historical inertia, to insist that the reverse set of conditions could not arise, given the appropriate political conditions (i.e., at a different place and time the system could go from 8 to 7 to 6, etc.).

We argue that this framework can be put to use to form an approximate scale, ranging from a pure Chayanovian peasant system to a pure modern, industrial/commercial agricultural system, with "modern peasants" or small-scale farmers as intermediate stages. This scale can be thought of as the degree to which the multiple balances of the Chayanovian system have been replaced by the simple accounting of gross returns (and associated modern rules such as long-term amortization, land speculation, etc.). While difficult to formally quantify, it is nevertheless clear that this can be conceptually represented as a gradient. Farmers' decision-making processes can then be located within some restricted range on this gradient, as some fraction of Chayanovian versus modern industrial.

The analysis of the transition from one ecological regime to another usually frames the issue as two alternative regimes (e.g., high biodiversity/low biodiversity; forest/savannah) emerging from a set of orderable conditions (e.g., annual precipitation), where the regimes themselves need not occur as alternatives (e.g., amount of tree cover, rather than the dual alternative forest versus savannah), but emerge from the underlying dynamics as alternative states. There is no reason to suspect that agroecosystems, as socioecological systems, need be exempt from this possible response. Indeed from the presence of alternative syndromes at the same point in space and time (e.g., shade and sun coffee in Mesoamerica, gardens and lawns in US suburbia, apple orchards and grain fields in Michigan, peasant farms and soybean plantations in Brazil), the strong suggestion is that some sort of hysteretic response to conditions may be in effect.[22] Yet there is no particular reason that only two alternatives exist. Indeed in theoretical ecology, simple systems frequently generate multiple stable states.[23] Thus, for example, if in a region where traditional agroforestry systems tend to be replaced by cattle pastures which in turn tend to be replaced by grain monocultures, it is not necessarily the case that such a progression will take place in an orderly fashion as a continuous change as sociopolitical conditions on a farm plot slowly change. Rather the three syndromes could be alternative stable states with tipping points between them (Figure 9.9).

One of the more interesting aspects of this framework is the situation presented in Figure 9.9c. Imagine that the upper part of the curve is the farm response (agricultural regime) of an impoverished agricultural system with low yields and unsustainable social factors (low wages for farmworkers, impoverished local communities, etc.), the bottom part of the curve is an industrial agricultural system with high yields but also unsustainable social factors, and a middle part of the curve represents an acceptable yield level with high sustainability. The dynamics of the system suggest that it may be impossible to attain that middle goal and no matter how much the conditions of production are changed, either moving to the right or moving to the left, the system will always either stay constant, either on the upper leg of the function or the lower leg, or jump from the upper to the lower or from the lower to the upper, always skipping over the desirable outcome. Such a structure has many implications. If desire is high enough (some would say the "political will" needs to be there also), knowing that simply pushing for normal changes in the conditions of productions the implications may seem daunting and suggest "surrender" to what might seem the forces of nature. But we suggest that the teachable moment here is the realization that what must be pursued is a change in the system itself. That is, if "changes in conditions" will never get to the desired state, the only way to get there is with a "change in system" program.

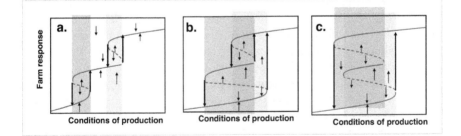

Figure 9.9 The farm response emerging from the conditions of production framed as three distinct syndromes of production, that is, three alternative stable states. a. Multiple (three) stable states for the farm response, where small hysteretic zones do not overlap. b. Multiple (three) stable states for the farm response, where moving the conditions in one direction leads to a pattern qualitatively distinct from moving it in the other direction. c. Multiple stable states for the farm response, but the interior state is effectively unattainable (hidden) once either the high or low state is reached.

Basin boundary collisions: chaos and catastrophe

The idea of the critical transition and tipping points as a model of surprise has become relatively standard in theoretical ecology. Another formulation, less well-known, derives from a slightly different point of view, involving the dynamics of chaos. Recalling the discussion of Chapter 4, chaotic attractors have their boundaries. While the variables of concern may change over time in a relatively unpredictable way, the limits of where they can exist are usually fixed. So a chaotic system may be completely unpredictable in a narrow region, but the overall attractor will normally have very fixed boundaries that will never be transcended. In addition to the boundaries, any chaotic attractor has a range of starting points that result in trajectories that eventually wind up on the attractor. This collection of points is formally denoted the "basin of attraction," metaphorically associated with the more physical basins of watersheds: rainfall is channeled from the mountainside to the river in the valley, the river being the attractor and the river basin being the basin of attraction (refer to Chapter 2).

Combining the realities of chaos and its boundaries along with the idea of a basin of attraction, some interesting phenomena can emerge. In Figure 9.10a we present time series for two alternative attractors; both are from the same model system representing a population under density dependent control. If the system is initiated near a population density of 8 or 9 individuals, it will always be constrained within 8 and 9 individuals approximately. It is chaotic, but, as we discussed in Chapter 4, there are clear limits on what values it might attain since the attractor has boundaries of 8 and 9. However, if the population is initiated between 1 and 7 individuals, it will always be constrained within about 1 and 7 individuals, chaotic to be sure, but never transcending the fixed boundaries of 1 and 7.

But now let us ask what happens if we slowly change the underlying parameters in the system. As the parameter changes over time (in this case we presume there is a parameter that affects the base line reproductive value of the population, it could be temperature or moisture, for example), the boundaries of both attractors slowly change. In addition to the changing boundaries of the two alternative attractors, there exist the edges of the basins. That is, each of the attractors exists within specific basins of attraction, regions in which all trajectories will eventually approach the attractors (a point that is far more obvious if the attractors are not chaotic, as discussed in Chapter 4). Note how the boundaries of the upper attractor slowly expand (Figure 9.10b) until they intersect the edge of the basin of the lower attractor, in what is logically called a basin boundary collision since it is the *boundary* of the chaotic attractor that overlaps the edge of the *basin* of attraction. What emerges is a perhaps surprising, sudden change in the behavior of the system (Figure 9.10b). Clearly, this is a completely distinct mechanism whereby ecosystem change can occur suddenly, and sometimes by surprise.[24]

Returning to themes introduced in previous chapters, here we see how the emergence of a relatively complicated pattern, like the one shown in the regions

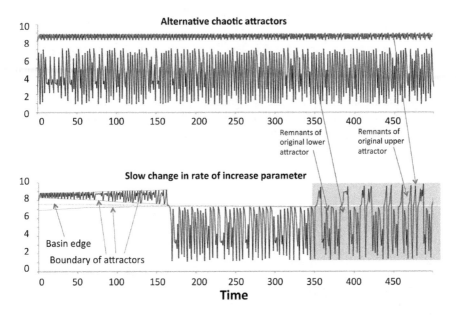

Figure 9.10 Time series trajectory of chaotic attractors. a. Two alternative chaotic attractors plotted on the same graph. The upper trajectory is what emerges when the system begins between about 8 and 9, and the bottom trajectory is what emerges when the system begins between about 2 and 6. b. With gradual change in one of the parameters, the lower boundary of the upper chaotic attractor intersects the edge of the basin of attraction of the lower attractor, making the upper attractor effectively disappear (i.e., it is no longer actually an attractor, formally). Yet with further change in the parameters, the lower boundary effectively intersects with what had been the edge of the basin of the upper attractor, pushing the system into the zone of the original upper attractor, giving rise to a new chaotic attractor. Note that the remnants of the original chaotic attractors can be visualized (emphasized with shading).

emphasized toward the end of the time series in Figure 9.10b, can be partially understood as the composition of two distinct attractive foci. Recall one of the messages of Chapter 4 that in the age of chaos, we may have to change our focus of studying ecosystems, asking questions about the morphology of attractors, rather than specific questions about particular trajectories, a theme repeated in Chapter 6 with the system of coupled oscillators of a disease and predator simultaneously exerting control over a pest.

Summary

The many evident cases in the past where major changes in some aspects of an ecosystem have happened suddenly and unexpectedly can be potentially explained by some recent elaborations of theoretical ecology. In particular, the idea of critical transitions, where various ecological regimes can be understood as flipping from one to another, can be thought of as corresponding to the general theory of critical transitions, a theory which is essentially equivalent to the mathematical theory of catastrophes. In this framing, there are zones of some particular measurement or parameter in which alternative equilibrium points exist, alternative ecological regimes, referred to as hysteretic zones. Changing that parameter may have very little effect on the ecological regime over a relatively large range, suddenly resulting in a dramatic transition of one regime into another. This way of framing the problem is one way of shedding light on the issue of unpredictability in ecological systems.

Notes

1 The classic work on this topic was by the Hungarian mathematician René Thom in his 1989 book.
2 Rival, 2005, 2012.
3 Perfecto and Vandermeer, 2015a.
4 The literature on this subject is now quite vast. Useful summaries are contained, for example, in Vandermeer and Perfecto, 1995; Swift et al., 1996, 2004; Gómez-Pompa, 1997; Altieri, 1999; Jackson, 2002; Tscharntke et al., 2005; Fischer et al., 2008; Padoch and Pinedo-Vasquez, 2010; Perfecto and Vandermeer, 2015a.
5 Our usage here is inspired by the anthropological literature on the history of agriculture, whereby there is a distinct trend that can be observed from hunting and gathering to simple cropping systems, to irrigation and terracing, ever increasing the intensity of the use of the land. Our use is perhaps less evident, but we think makes sense as a rough metaphor for the way agriculture changes in the contemporary world.
6 Carpenter et al., 2011.
7 Staver et al., 2011a, 2011b; Hirota et al., 2011.
8 Leterme et al., 2006.
9 Although at the time of writing there has emerged a burgeoning literature on the anticipation of these critical transitions. See, for example, Scheffer et al., 2012.
10 Staver, Archibald and Levin, 2011; Hirota et al., 2011.
11 Hirota et al., 2011.
12 McCook and Vandermeer, 2015; McCook, 2006; Avelino et al., 2015.
13 Perfecto and Vandermeer, 2015a.
14 Jackson et al., 2012; Vandermeer et al., 2014.
15 Vandermeer and Rohani, 2014.
16 These complications are discussed in various disjoint publications, namely, Vandermeer et al., 2010, 2014, 2015; Jackson et al., 2012, 2012a; Perfecto et al., 2014.
17 Vandermeer et al., 2015.
18 Andow & Hidaka, 1989; Vandermeer. 1990.
19 Moguel and Toledo, 1999; Perfecto and Vandermeer, 2015a.
20 Van der Ploeg, 2009, 2013; Friedman, 1980; Ellis, 1993; Bernstein, 2009.
21 In our view the best recent summary of the Chayanovian system is by Van der Ploeg, 2013.
22 Vandermeer and Perfecto, 2012.
23 Beisner et al., 2003.
24 Vandermeer and Yodzis, 1999.

10 The "scientific" basis of agroecology

The underlying framework of this book is based on two assumptions. First, we suggest that complexity in its modern sense can be thought of as a backbone of today's ecological science. Second, we argue that ecology should be the foundational science of agroecosystems. Taken together, we are compelled to conclude that ecological complexity should be the scientific foundation for agroecology going forward.

Although the science of complexity has multiple touch points with many disciplines, and encompasses a rather large range of generalized topics, our reading of the literature through an agroecological lens suggests that there are seven relevant and interconnecting themes – Turing effects and spatial structure, chaos, stochastic forces, coupled oscillators, multidimensionality, trait-mediated interactions and critical transitions. These seven themes form the core of this book. What is perhaps less obvious from our narrative is the fact that these themes intersect in important ways to form something of a more generalized framing. Throughout, we have made reference to points of contact among them, but devising a more general statement about their intersection is a somewhat more difficult task. We could invent particular fantasies in which all seven themes are interlocking (e.g., a trait-mediated interaction added to a system of coupled consumer/resource pairs generates chaos, which expands to a basin boundary critical transition, forming spatial patterns which themselves are sustained by essential intransitivities, which allow for multiple coexistence of a high-dimensional system). Indeed, such fantasies may turn out to be a bit less Panglossian than might first seem to be the case, the result of ecosystems being complex.

Specific intersections among these seven themes, many of which we cited throughout the various chapters, are likely to characterize real problems that emerge in agroecosystems. Thus, for example, as we noted in Chapter 6, when a potential pest species is attacked by two different biological control agents, the underlying structure of that system is certainly one of coupled oscillators and might very well be chaotic, with the important practical result that each of the biological control agents separately is ineffective, but taken together may function well, and furthermore, that result emerges from the complicated issue of intersecting chaotic attractors. To take another example, we noted in Chapters 8 and 9 that a critical transition to epidemic status of a crop disease may result from multiple indirect interactions, some of which are characteristically trait-mediated.

Yet the literature is filled with other examples that, we believe, reinforce the idea of complexity in the agroecosystem. For example, combining the chaotic forces of farmers making independent decisions about planting based on incomplete economic information (as noted in Chapter 4), if combined with the ecological forces associated with landscape density dependence, result in the potential for basin boundary collisions and ultimate surprising behavior of price and production trajectories,[1] the combination of chaos and critical transitions. Or, the existence of intransitive loops generates particular spatial patterns[2] that may form a variable habitat structure in which other organisms in the system are forced to live.[3] Many other potential structures could be cited, echoing the admonition of Gary Polis[4] many years ago about the evident complexity of ecological systems. Our intent is to place these seven theoretical themes of complexity as potential tools for understanding the real complexity of agroecosystems.

Having argued our vision in the last seven chapters, we now reflect on both the idea of complexity as something special for the science of ecology, and then on the structure of agroecology as we see it going forward, through this particular lens.

Complexity science and ecology

There has been a great deal written in both technical and popular outlets about the complex nature of ecological systems. The famous spaghetti graph of the codfish food web presented in Chapter 1 is an early example (Figure 1.1). Its effect has been to interject a degree of humility into narratives about how far the science of ecology has progressed.

While calls to acknowledge the many connections that exist among a multitude of ecological elements are clearly sensible, they represent only a small section of what we propose is the true complexity of ecological systems. That complexity is rooted in a variety of surprising, sometimes astounding, results from mathematics, from crucial experiments in ecology, and from continuing and irreplaceable natural history observations. It is a framework that includes the somewhat unexpected conclusions that complex patterns emerge from very simple rules while at the same time simple patterns emerge from complex rules. It is rooted in fundamental principles of western science, while acknowledging insights from what earlier writers like Boyle and Darwin referred to as "the trades." As we wrote in an earlier summary of pest control in the coffee agroecosystem,

> In the end, [we suggest] that the vision of the natural world as harmonious and balanced is wrong if we naively accept an unreconstructed Newtonian world view of balance – ecosystems are not like a marble coming to rest at the bottom of an inverted cone. However, through the spatially explicit complexity of myriad interactions, many of which are multiply nonlinearly,

a higher notion of balance emerges – not the balance of Newton, but rather the balance of a shifting sand dune whose detailed structure changes minute to minute, but whose fundamental nature as a "sand dune" is never in doubt. Our understanding becomes not the crude, positivist logic that must identify a singular enemy to conquer, and a magic bullet with which to do so, but rather the holistic vision of a new kind of "balance" emerging from the very complexity that traditional farmers intuitively understood from the beginning.[5]

That is, we postulate a convergence of diverse traditional knowledge systems with modern Western models of complexity, into a dialectical whole, envisioning agroecosystems as complex systems.

The transformation in understanding that we feel is emerging generally from an appreciation of the reality of complexity, reminds us of some earlier transformations. For example, building on ancient conjectures of Greek thinkers such as Democratus and Epicurus, in 1623 Galileo Galilei said:

> Philosophy is written in this grand book, the universe, which stands continually open to our gaze. But the book cannot be understood unless one first learns to comprehend the language and read the letters in which it is composed. It is written in the language of mathematics, and its characters are triangles, circles, and other geometric figures without which it is humanly impossible to understand a single word of it.[6]

What Galileo meant was far more literal than a modern reader might expect, thinking literally of solid objects like triangles and circles combining in many ways to create everything in the universe. This sort of thinking continued evolving, eventually giving way to a more systematic understanding of the particulate nature of the universe, the world of atoms (not that much of a change when we realize that Galileo's triangles could be very, very tiny) and their changes in space and time. This viewpoint culminated, philosophically, in 1814 with Laplace's Demon, an entity described as:

> An intellect which at a certain moment would know all forces that set nature in motion, and all positions of all items of which nature is composed, if this intellect were also vast enough to submit these data to analysis, it would embrace in a single formula the movements of the greatest bodies of the universe and those of the tiniest atom; for such an intellect nothing would be uncertain and the future just like the past would be present before its eyes.[7]

This point of view says that the present, as we know it, is a consequence of the past, but that in the same sense, the future is a consequence of the present, which, if we know with sufficient precision and have the computational power at our disposal, will enable us to predict that future.

In a sense, it is this philosophical principle that the science of complexity negates, almost as dramatically as the atomic theory of matter negated Galileo's solids theory. True, the entire enterprise of statistical mechanics arguably dealt it a near-fatal blow with its astounding predictive ability based on probability theory as opposed to Laplace's deterministic assumptions.[8] But its spirit is challenged in a more practical and enlightening way by the burgeoning literature in complexity science that we see today. Applied to the science of ecology, speaking generally, complexity science provides us with a new paradigm for viewing ecological systems. Their patterns in space, their chaotic characteristics, their basic trophic oscillations, their multidimensionality, their non-linearities, their fundamental stochasticity and their tendency toward critical transitions are the issues that make up that new paradigm. Or, rather, it is the intersection of these ideas that do so. When applied to agroecology, it is a paradigm that provides a springboard to apply ecology more deeply to the understanding of agriculture.

As we have constructed this book, we have taken our clues from both the natural world (both actively managed agroecosystems and currently unmanaged ones), and from the ongoing development of complexity theory. We see a strong connection between the two. The seven themes of complexity we put forth are common themes in the literature on complexity science (not by any means all-inclusive) and thus are justifiably referred to as seven themes of complexity. This is, of course, a social construction. A chaotic system is well-defined, as is a stochastic system, and critical transitions and so forth. But the essence of our argument is implicitly deeper than that. Because of the complicated nature of ecosystems in general, because of their basic multidimensional nature, their chaotic and stochastic characteristics, the indirect interactions contained therein, their nature as coupled oscillators, and their tendencies toward critical transitions and self-organized spatial patterns, we argue that they are generally complex systems. In other words, the socially constructed science of complexity seems parallel to the way natural ecosystems operate.

The four pillars of agroecology

As agroecology continues its evolution as a multidisciplinary topic combining many traditions, both popular and academic, we sense the slow emergence of a consensus on something of a megastructure. It is possible to recognize four, sometimes competing but mainly converging lines of thought that in a classical Marxist sense seem to be "interpenetrating." Science, traditional knowledge, political resistance and the natural world, are, we propose, the four pillars supporting the emerging platform of agroecology.

The background in formal science

It is a core argument in this book that the science of ecology should be the foundation of the new agriculture, a claim that is certainly not new with us. However, our vision is somewhat unorthodox in two ways – first in our self-conscious critical vision of science itself, and second in the related issue of what the science of ecology is and ought to be. At this point, further reflection on these issues is warranted, first on the nature of science itself, adding to the preliminary reflections we presented in Chapter 1, and second on the philosophical nature of the science of ecology as practiced in the academy today.

When Galileo trained his telescope on Jupiter, he saw the moons. Several sequential drawings of the position of those moons led him to imagine that they were orbiting around the far larger mass of the planet Jupiter.[9] He deduced that they were nothing special, that as those moons orbited the planet, the planets orbited the sun, which Copernicus had deduced earlier, a deduction that got Galileo confined under house arrest for the best part of his later years. A half century later Newton noted:

> a leaden ball projected from the top of a mountain . . . parallel to the horizon . . . is carried in a curved line to [some distance] . . . [by] increasing the velocity we may at pleasure increase the distance to which it might be projected, and diminish the curvature of the line, which it might describe, till at last it should fall at [a greater distance] . . . or even might go quite round the whole earth before it falls [or even] so that it might never fall to the earth.

Thus from the simple observation that "gravity is the thing that attracts masses to one another" and the practical observation that almost anyone could easily imagine (the leaden ball), Newton envisioned the generation of a mass orbiting another mass and deduced that the orbits Galileo and Copernicus had seen were a simple consequence of this "law," which today we know as the law of gravity. The fact that there was some simple mathematics that could be applied to the system should be viewed as a technical footnote to the main insight of a lead ball propelled parallel to the surface of the earth at great velocity. And, most importantly, Newton had an underlying ideology that provided something of a moral compass to his work – he was seeking to understand the mind of God!

Newton was to the science of physics what Darwin was to the science of biology, but here again we see an important element of ideology involved. Most elementary biology texts treat Darwin as a loving father, in love with nature and biodiversity almost as much as with his wife and children, passionate about understanding where the biodiversity he loved so much came from. He fits nicely with the standard narrative about science – the lone genius, passionate about his subject, seeing further and deeper than anyone before him. But a less romantic and more political narrative has recently emerged. The two curators of Darwin's letters at Cambridge, Adrian Desmond and Robert Moore, wrote what seems to us an eye-opening book, *Darwin's Sacred Cause*, in which they detail how Darwin's commitment to the worldwide anti-slavery movement propelled him to finish his masterpiece. The argument is complex, but to us, quite convincing.[10] For our contention we simply note that again, an icon of science had important ideological baggage that undergirded his science.

More to the direct subject matter of this text, one of the founders of modern ecology, Arthur Tansley, was a vigorous defender of the post-WWI imperial efforts of the British empire in Africa, especially in South Africa. Standard ecological narratives simply note that Tansley provided us with the concept of the ecosystem, sort of as an alternative to Clements' idealistic notion of the ecological community as an organism. While Tansley's materialism is rightly celebrated, his motives were perhaps less noble, effectively seeking to provide Great Britain with a scientific foundation for the organization of colonial Africa, including the notion that the native people needed to be managed, sort of like the wildlife. The details are spelled out in the excellent *Imperial Ecology* by Peder Anker.[11] Again, the arguments are complex and could certainly be contested, but for our purposes we note that the science of ecology, like any other science, has always been influenced by the sociopolitical system in which it is embedded. Whether acknowledged explicitly or not, it has had strong ideological underpinnings ever since its beginning, and it is not surprising that such dynamically changing ideologies continue their influence, however obscure they may seem to the scientists themselves.

Nevertheless, the future of agroecology depends on the growth of scientific understanding of how agroecosystems work. While much that we can commend to the practice of agroecology is in the form of "rules of thumb" (monocultures are generally not a good idea, compost is excellent fertilizer, biological control can be aided by planting flowers next to the crops, and many others), much that we hope to be able to do in the future depends on furthering our understanding, scientifically, of the ecology of agroecosystems. We advocate for the encouragement of scientifically oriented agroecological promoters to take seriously advanced training in sciences such as chemistry, and plant physiology, but most importantly, modern ecology and mathematics.

Science and the scientific method have become foundational in the modern world. Although "science" is frequently used and abused to support nefarious ends, the basic principles of science remain a powerful way of understanding those parts of the natural world that are amenable to its application. It is evident that there are myriad "ways of knowing" and it would be foolish to suggest that some of them were illegitimate *a priori*. However, when it comes to organizing human interference in the natural world, the use of classical scientific principles, whether in the construction of an airplane or an agroecosystem, is essential. Rejecting science, for whatever reason, burdens us with trying to solve problems with manacled hands. Worse, in contested terrain (e.g., the debate over GMOs, current as of this writing), the side that adopts an anti-science position has a huge handicap. Noam Chomsky famously introduced the notion of "intellectual self-defense" into the political lexicon. A great deal of nonsense pervades our world, and that nonsense is frequently the basis for exploitation, the proverbial attack of the snake-oil salesperson. Having a scientific understanding about how nature works, and understanding what constitutes scientific evidence, provides one with a potential defense against such attacks. As the agroecological revolution spreads, there is nothing preventing quackery, intentional or accidental, to enter the revolutionary process. Understanding the basics of ecology (and chemistry, physics, mathematics, etc.) can contribute to that intellectual self-defense.

Traditional knowledge

Anthropology has documented, sometimes in great detail, how local and regional traditions have been incorporated into agricultural practice in complex and sometimes enigmatic ways, providing an important window into the impact of agricultural practices on cultural development and change. Yet the modernizing paradigm sort of reverses this understanding – development dictates which of those traditional practices deserve to be retained (as it turns out, hardly any). Consequently, traditional agricultural practices have been dramatically marginalized as the industrial agricultural system has taken hold worldwide. Thrown in the ashbin of history, these traditions are thought at best as irrelevant to modernization, at worst an impediment to the triumphant industrial agricultural system.

We join many others to claim that this perspective is detrimental to the goal of developing a just and sustainable food and agricultural system, in short the agroecological approach. As Altieri noted years ago, traditional practices represent important waypoints in the route to developing that new agriculture. As the technical aspects of those practices begin to get written down, systematized, and incorporated into "Western Science," they can be viewed less as candidates for the ashbin and more as living systems that, understood with the tools of science, can be integrated among themselves into a new paradigm. A new breed of technical expert, perhaps better classified as technical chroniclers, seeking to catalogue and understand each of the incredibly diverse traditional practices around the world. This is clearly part of the recipe we need for the new agriculture.

Foremost among those new technological chroniclers were Alfred Howard and Gabriel Matthaei, a husband and wife couple, both British botanists, as we noted in Chapter 1. The Howard/Matthaei team was dispatched to colonial India to teach Indian farmers the modern techniques of agriculture. Rather than doing that, they observed and learned from traditional Indian farmers and, with that knowledge, developed a variety of techniques that would today be called organic. What they learned from local traditional farmers lead them to reject, at least in part, the growing industrial system.

While the Howard/Matthaei team is perhaps the most noted in this discourse, others have also contributed in significant ways. For example, Franklin Hiram King, a soil scientist from the University of Wisconsin, in 1909 traveled through Asia documenting traditional farming practices. His book *Farmers of Forty Centuries: Organic Farming in China, Korea and Japan* has become a classic in the organic agriculture literature. Today's agroecology has also been greatly influenced by documentations of pre-Hispanic agriculture in the Americas, starting with Cook's 1916 article in *National Geographic*, "Staircase of the Ancients," that documented the *chinampas* of the Aztecs, continuing with Latham's 1936 *La Agricultura Precolombina en Chile y los Paises Vecinos (Pre-Colombian Agriculture in Chile and Neighboring Countries)*, and later with the vast work of Efrain

Hernandez Xolocotzi, who contributed to the biological and ecological foundations of agroecology from his studies of traditional systems in Mexico.[12]

Ultimately, the importance of traditional knowledge stems not only from the
experiences of these early observers, and the continuing experiences of today's
practitioners, but from a deep sense of how agricultural practices must be tied to
the land and that those having survived for many generations would have gone
through a sort of natural selection process. To suggest that traditional systems
are always good would be folly since many traditional systems extinguished
themselves,[13] but as a general rule, systems that have been around for a long
time must have some component of sustainability associated with them.[14] The
modern industrial system, on the contrary, is young and, based on its necessary
short-term framing in terms of profit maximization, cannot be expected to
incorporate concerns other than monetary in its developing discourse. Thus, we
see both an empirical and theoretical reason to take traditional practices as "raw
material" to feed into the ongoing development of agroecology.

Political resistance

Peasant revolts are common throughout history, fueled mostly by social injustices
and struggles for autonomy and the right to work the land. Historical records
embrace at least 2,000 years, from the peasant revolt against the Qin dynasty in
209 BC that led to the fall of the dynasty, to the anti-NAFTA Zapatista Revolution in Chiapas, Mexico, in 1994. In Europe we can trace important influences
from the English Revolution of 1640, in which peasant farmers changed the
world through demands that today would be referred to as sovereignty (it also
involved a king having his head cut off), leading to a degree of emancipation.
Subsequently the peasant class became again marginalized, and by 1649 their
situation was reverting to the former conditions of serfdom. At that point a
movement known the "Diggers" emerged with their radical agrarian program as
we briefly described in Chapter 1. The Diggers may have been short-lived, but
their ideology lives on as a militant attachment to the land and the right to use
it, and a rejection of illegitimate power, especially a power that cynically abuses
and sometimes destroys the "Common Land," or in modern parlance, the natural
world. Perhaps the "Diggers" and their kindred were the first agroecologists.

This continuing tradition is, in our view, a major component of the movement known as agroecology – social organization and political action. In Brazil,
for example, the MST (*Movimento dos Trabalhadores Rurais Sem Terra*, or Landless
Workers Movement) is one of the largest rural social movements in the world,
organizing rural workers to "take back the land." They organize groups of rural
families who have no land and are among the world's most marginalized people.
Targeting land that is, with questionable legitimacy, held by large landowners, the
MST helps groups of families (sometimes hundreds at a time) to occupy such
land, with the ultimate goal of wresting title of the land to the occupiers, eventually forming small farming units. Although the MST began as a rural social and
political movement with an almost exclusive focus on land acquisition and little

focus on the technology of agricultural production,[15] through internal analyses and political discussions the movement has come to embrace agroecology as a way of generating more autonomy for the small-farming sector.[16] The history is still being written, but the successes of the movement have been felt in Brazil for decades.

Even more expansive at the present time is the umbrella organization known as La Via Campesina (LVC – "the farmer's way," in Spanish). LVC emerged as a movement opposing neoliberal globalization that affected peasant farmers the world over and has had a major impact on rural political activity throughout the world. This movement brings together farmers, fisherfolk, and rural workers from around the world in "commission meetings," where major issues of interest to rural communities are discussed, political actions considered and planned, and basic philosophical positions developed democratically. In our view, LVC carries the spirit of the Diggers to the present day. More importantly for the topic of this book is LVC's adoption of agroecology as a tool of resistance that helps farmers gain more autonomy from external markets and enhance their agricultural systems for the future.[17]

Messages from the natural world

Anthropologists rightly emphasize that pre-industrial indigenous people not only had an appreciation and knowledge of local ecological forces, they applied that knowledge to construct their agricultural systems. From the Mexican *chinampas*[18] to the Xinghua's *duotian* raised fields,[19] from the *milpas* of Mesoamerica to the *kihamba* agroforests of Tanzania, from the Javanese home gardens[20] to the *tapade* system in Guinea,[21] technological inputs into agricultural systems bore the stamp of intimate knowledge of local ecological principles. Although the articulation of those local ecological principles may not have been recognizable in modern ecological terms, and frequently took on the semi-religious form of veneration of "mother earth," the close association with the natural world that inevitably was the case with early farmers required them to become ecologists in principle. Although their knowledge may have been geographically restricted, in that it was based strictly on the local landscapes that they experienced, it was nevertheless deep and, most importantly, was based on their understanding of how the natural world works.

In more recent times, and in the same spirit, we see the important influence of Richard St. Barbe Baker, who served the British empire in Africa and developed a deep and effectively spiritual tie to forests. As a formally trained forester he was sent to Kenya where he found a scarred landscape, the result of a century of irresponsible farming under British rule. He set out to reforest large sections of the country, a task for which he formed the "Men of the Trees" movement (today's International Tree Foundation).[22] Much of interest is to be found in Baker's writings about forests, but for purposes of this text, he saw in the forests lessons that would become important for agroecology (e.g., natural cycling of nutrients occurred in natural forests but not in many of the agricultural systems that displaced them). He sought insights from "natural" ecosystems that would guide the planning of agriculture, a forerunner for what would become "natural systems agriculture."[23]

Today it is common to assert, as many ecologists have done, that nature provides a template on which planning of all sorts should be based.[24] From the forest-like environment of the shaded coffee system, the base of many examples in this book, to the promotion of scattered trees in pastures, mimicking the basic structure of a tropical savannah, natural terrestrial ecosystems are taken, sometimes with very little fanfare, to be models on which agricultural planning should be (and sometimes is) based. Most influential in this framing is the perennial polyculture program of the Land Institute, in Kansas, originally promoted by the geneticist Wes Jackson. Jackson noted that native temperate grasslands are generally composed of perennial polycultures, while the massive grain-producing commodity production of much of the industrialized system has converted this highly productive (and, we add, rich in carbon storage) natural system to a hopelessly degraded annual monoculture. Using natural systems as a template, the Land Institute generated this new and hopeful framing of perennial grain-based agriculture.

Perhaps more important are the details of how particular modalities function in natural ecosystems, and the likelihood that they will function similarly in the agroecosystem context. For example, hemipteran bugs are frequently agricultural pests that suck juices out of the crops, and they are frequently protected from their natural enemies by ants that harvest the sweet honeydew they produce. It is thus logical to presume that the modality ant-hemipteran is likely to have a negative effect in the agroecosystem. However, there is substantial evidence from natural systems that plants actually benefit from the association since the ants frequently deter or prey on other herbivores, thus exerting a positive indirect influence on the plant. In our work in Mexico we discovered precisely this mechanism at work with an ant and its mutualist, the green coffee scale. Although the ant is a "pest" in the sense that it protects the scale insects, it also acts to deter an even more devastating pest, the coffee berry borer, forming a modality in the agroecosystem that is well-known and easily recognized in natural systems.[25] The natural system mode of ants tending hemipterans has been effectively integrated into the agroecosystem, in this case with little planning.

The "whole" of agroecology

Agroecology, in our view, lies at the intersection of these four components: (1) a strong scientific foundation that provides a broad understanding of the ecological principles that underlie agroecosystems; (2) traditional and local knowledge that has been accumulated through generations of trial and error and that provides deep understanding of practices that work under very specific conditions; (3) social/political organization to push forward an emancipatory agenda for the agroecological practitioners (farmers), and (4) the local natural systems that provide underlying clues to both scientists and farmers. These four components form the essential pillars upon which the agroecological platform should be based. Although we understand that other framings could be imagined, and we admit that the issue is far more complicated than the four-dimensional version we offer here, this simplified framing provides us with a readily understandable philosophical foundation and enables us and other agroecological practitioners/ scientists to see connections among diverse activities.

But we hasten to acknowledge that formal science's contribution to agroecology is trivial, perhaps even illegitimate, if viewed in a vacuum. Only in the context as stated here do the advances of Western-style science attain true relevance. Our argument, spelled out in Chapter 1 of this book, is that the science of ecology should be the foundational science for the development of agroecology (again, only as one of the four pillars). And we have occasionally lamented that the popular version of ecology applied to agroecology has little to do with what has been happening in the science of ecology over the past 50 years (e.g., for some especially recent converts to the agroecological movement, ecology seems to be "that which I learned from my parents on backpacking trips and from the Discovery Channel," which is what a recent student told us). As we noted in Chapter 1, Rachel Carson provided us with a set of observations that strongly emphasized interconnections and interactions, with a vision of complexity that qualitatively mirrors what later developed as ecological complexity. Our framing takes a cue from Carson, but is elaborated in the light of qualitative advances in ecological modes of thought over the past half century. It is a vision of complexity that coincides with what Richard Levins had advocated for the past 50 years and with what much of the current wave of complexity science has been discovering (or constructing?).

Solving farmers' problems in the age of ecological complexity

At a practical level, farmers the world over have problems. And problems, almost by definition, seek solutions. Many agricultural experiment stations repeat the mantra that their fundamental goal is to solve the problems that farmers face. It is worth repeating the fundamental principle we have told several times earlier, elaborated so eloquently by Guatemalan-Mexican agroecologist Helda Morales (let's call it the Morales effect).[26] The fundamental goal should not be to "solve the problems that farmers face," but to design systems that minimize problems in the first place. And, as we argue in this book, designing systems that obey rules that are only partially understood is difficult, to say the least. Thus, a major research agenda should be simply to understand the complex ecology that exists in all agroecosystems. Of most importance is understanding how small-scale, more traditional type agroecosystems work, how they are similar to or different from natural systems surrounding them, how they connect with sociopolitical forces surrounding them. Furthermore, as is a well-known mantra in pure ecology, most ecological interactions unfurl in the context of the larger landscape within which any particular farm is embedded.

Our insistence on understanding ecological complexity will frequently require answering the farmer's query, "what can I do?" with the honest answer, "I do not know." We recall a farmer we interviewed in Costa Rica many years ago. We asked him what technical researchers should emphasize. We were expecting some priority list, such as soil fertility, or pest control, or market access. But his answer was simple and more profound. He simply said "don't lie to the farmer." To suggest a formulation of bacteria and fungi that will enhance soil fertility, without really knowing whether it works or not under those particular conditions, is a form of lying to the farmer. To suggest a pesticide without a background of experiments showing that it does what it claims it does, or to assure her that it will not harm the health of her family, is lying to the farmer. As much as it is troubling to the farmer to face seemingly intractable problems on the farm, it is troubling to the technical expert, whether in the industrial or agroecological framing, to have to say "I don't know what you should do." Yet, if the principle of not lying to the farmer is accepted, that answer is frequently the correct one.

There are, however, cases in which the answer to the farmer's query is obvious but unwanted. An example may be the recent explosion of the coffee rust disease where we do our research, in southern Mexico and extending throughout Central America all the way to Peru. Our ecological analysis suggests that the most important factors leading to the sudden epidemic of this disease are landscape factors associated with the loss of shade, both in the coffee itself and a more general landscape effect of deforestation. If we are to be honest with the farmer who asks us what to do, our answer should probably be "there is nothing you can do alone on your farm." And then, if we extend our honesty to true prescription, we should add "organize your neighbors and fellow citizens to promote reforestation, both of lands deforested for cattle production and deforested coffee farms." Naturally, this last recommendation will be thought to be political. But honesty should compel us to move in that direction.

It is our expectation that adopting the research agenda of ecological complexity will steer us away from giving recipes to farmers and allow for the development of general guiding principles, some of which we have indicated sporadically throughout the text. Much as we feel the history of technological developments in agriculture in the past have a mixed record, we expect that ecological complexity as a new framework, will also have mixed results. The tendency that remains in most applied research is to solve problems, to be sure, but searching for solutions is frequently taken to mean developing a recipe. Whether a technological "package" from an industrial agriculture field station or a set of rules based on traditional practices, recipes are frequently themselves recipes for failure. Well-meaning promoters sometimes seek simple recipes that vague experience suggest will work, in the absence of either hard evidence or theoretical reasoning. The idea of intellectual self-defense applies here. Our goal is to help with the ongoing development of understanding how the agroeco-system works with the concomitant wisdom that the farmer him and herself can apply to construct the farm in such a way that problems do not develop in the first place.

The importance of thought-intensive technology

We end as we began in Chapter 1, by noting the existence of two chief world agricultural systems. There is the world where technology, in its Western world-view sense, will solve all problems with a simple recipe, a product, and where science is important but takes second place to the need for maintaining the socioeconomic system, and moreover, is not in the hands of the farmers themselves. Then there is the alternative world where a different sort of technology, one based on understanding the ecological complexity of the system, is the way to solve problems in the long term. This alternative technology is based on true scientific understanding. And to the popular conception that technology will solve all problems, the new framework does indeed represent a technology. To use a phrase coined long ago by Richard Levins, it is "thought-intensive" technology. We hope that our framework of ecological complexity can be part of the toolbox for the promotion of this vision.

Notes

1 Vandermeer, 1990.
2 Vandermeer and Yitbarek, 2012.
3 Jackson et al., 2014.
4 Polis, 1991.
5 Vandermeer et al., 2010.
6 The quote is from Galileo's famous "The Assayer," encountered in Wikipedia at https://en.wikipedia.org/wiki/The_Assayer.
7 Laplace, *A Philosophical Essay on Probabilities*. Published sometime before 1827. English translation from the sixth French edition by F. W. Truscott and F. L. Amory in 1902 by John Wiley and Sons.
8 To be sure, there are major philosophical debates involved with this issue. We take it as a tentative and approximate story that quantum mechanics so deviates from the spirit of Laplace's Demon as to obliterate the idea. Other analysts are not so sure. See Binder, 2008.
9 Our home institution, the University of Michigan, has as one of its prized possessions a single leaf manuscript of Galileo's that is a draft of a letter to the Doge of Venice, describing the telescope he had built, emphasizing its potential military utility. Apparently he used this draft to sketch his nightly observations of the moons of Jupiter, clearly indicating with arrows how he thought those moons were orbiting their home planet. It represents a major moment in science. A copy of the letter and sketches can be seen at www.lib.umich.edu/special-collections-library/galileo-manuscript.
10 Desmond and Moore detail the long history of the Darwin family to the anti-slavery movement and report extensively on Darwin's musings of how horrible the institution was. Although Desmond and Moore did not directly state it, our interpretation from their book is that Darwin was of the mind that if he could show that all humans, like all organisms, are part of effectively the same family, that would undercut the underlying biological justification for slavery. If God did not simply make black people different for some particular reason, how come they have such a different skin color? It was just a matter of natural selection and our family tree extends to include all humans in the same family.

11 Peder Anker's *Imperial Ecology* lays out the basic ideas. In the immediate aftermath of WWI, Britain was riding high on its imperial mission. Understandably most Europeans at the time could not imagine that they might once again repeat the carnage of WWI and the assumption was that the basic imperial structures emergent post-WWI would be more or less permanent. It thus was important that Britan recognized its "responsibilities" as an imperial power. This meant that it needed to know what existed in its vast holdings so as to properly manage it. The previously planned botanical surveys of South Africa and elsewhere, organized through Oxford by Tansley became important not only for their aim of understanding the distribution of plants and animals (including people), but for understanding the order of nature itself. Thus the original botanical surveys became effectively ecological surveys, with the concomitant idea of an ecosystem emerging. Thus it is not only through the criticism of Clementsian idealism that Tansley conceived of the ecosystem as a material thing, but also through the influence of his organizing the botanical surveys of the colonies. Tansley's main adversaries were South African. A slightly extended version of what seems a similar thesis was provided by Helen Tilley in the self-descriptive title "Africa as a Living Laboratory: Empire, Development and the Problem of Scientific Knowledge."

12 Rosado-May, 2015; Gómez Pompa and Venegas, 1982.

13 The many traditional societies based on special agricultural techniques include the Hohokam of southwestern North America (McGuire, 1982), and areas of ancient Meso-potamia (Wilkinson, 1997), both of which seem to have ignored the evident problem of inevitable salinization from irrigation. Suggestions that collapse of traditional agricultural systems are also sometimes made for the Ancient Mayans (Demarest, 2004), although reviews by Tainter (1990, 2006) emphasize the complexity of social collapse and, most importantly, find little evidence that a Malthusian overshoot has ever been involved.

14 There is a very large literature claiming this point. For example, Altieri 1990; Altieri et al. 2012; Gliessman et al., 1981; Wilken, 1990.

15 Wright and Wolford, 2003.

16 Holt-Giménez, 2009.

17 Rosset and Martinez-Torres, 2013, 2014.

18 Delgadillo Polanco, 2009.

19 Si-ming, 2011.

20 Soemarwoto and Conway, 1992.

21 André et al., 2003.

22 St. Barbe Baker, 1948.

23 Wes Jackson, the founder of the Land Institute in Salina, Kansas, has been a vocal proponent of this idea. See, for example, Jackson, 2002.

24 Perfecto and Vandermeer, 2014; Dawson and Fry, 1998.

25 For the whole idea of modalities see Perfecto and Vandermeer, 2015b; for the effect of ant/hemipteran mutualisms on plants in general see Belsky, 1986. Also see the discussion in Chapter 8 of this particular example.

26 Interviewing Guatemalan peasant farmers, Morales began by asking "what pests do you have?," to which all the farmers answered, "we have no pests." She then changed the wording of the question to "what kinds of insects do you have in your corn?," to which the farmers gave her a large list. Comparing the list the farmers reported, Morales discovered that many of those insects were recorded as corn pests in the literature. Questioning the farmers as to why these insects, known to be pests elsewhere, never gained pest status in their farms, she basically got the answer that the management of the farm was such that the potential pests remained only potential (Morales and Perfecto, 2000).

References

Abrams, P. A. (2007). Defining and measuring the impact of dynamic traits on interspecific interactions. *Ecology*, *88*(10), 2555–2562.

Alexander, A. (2014). *Infinitesimal*. London: Oneworld.

Allen, D. N. (2012). *Feedback between ecological interaction and spatial pattern in a transitional Michigan forest*. PhD dissertation, University of Michigan.

Altieri, M. A. (1987). *Agroecology: the scientific basis of alternative agriculture*. Boulder: Westview Press.

Altieri, M. A. (1990). Why study traditional agriculture? In C. R. Carroll, J. H. Vandermeer, & P. Rossett (eds.), *Agroecology* (pp. 551–564). New York: McGraw-Hill.

Altieri, M. A. (1999). The ecological role of biodiversity in agroecosystems. *Agriculture, Ecosystems & Environment*, *74*(1), 19–31.

Altieri, M. A. (2004). Linking ecologists and traditional farmers in the search for sustainable agriculture. *Frontiers in Ecology and the Environment*, *2*(1), 35–42.

Altieri, M. A., Funes-Monzote, F. R., & Petersen, P. (2012). Agroecologically efficient agricultural systems for smallholder farmers: Contributions to food sovereignty. *Agronomy for Sustainable Development*, *32*(1), 1–13.

Andow, D. A., & Hidaka, K. (1989). Experimental natural history of sustainable agriculture: syndromes of production. *Agriculture, ecosystems & environment*, *27*(1–4), 447–462.

André, V., Pestaña, G., & Rossi, G. (2003). Foreign representations and local realities: Agropastoralism and environmental issues in the Fouta Djalon tablelands, Republic of Guinea. *Mountain Research and Development*, *23*(2), 149–155.

Anker, P. (2009). *Imperial ecology*. Cambridge, MA: Harvard University Press.

Armstrong, R. A., & McGehee, R. (1980). Competitive exclusion. *American Naturalist*, *115*(2), 151–170.

Avelino, J., Cristancho, M., Georgiou, S., Imbach, P., Aguilar, L., Bornemann, G., . . . & Morales, C. (2015). The coffee rust crises in Colombia and Central America (2008–2013): Impacts, plausible causes and proposed solutions. *Food Security*, *7*, 303–321.

Bach, C. E. (1980). Effects of plant density and diversity on the population dynamics of a specialist herbivore, the striped cucumber beetle, *Acalymma vittata* (Fab). *Ecology*, *61*(6), 1515–1530.

Bairey, E., Kelsic, E. D., & Kishony, R. (2016). High-order species interactions shape ecosystem diversity. *Nature Communications*. http://dx.doi.org/10.1038/ncomms12285.

Bak, P. (1996). *How nature works*. New York: Springer.

Balfour, L. (1982). *Towards a Sustainable Agriculture the Living Soil*. Soil Association.

Bassett, D. S., & Bullmore, E. T. (2016). Small-world brain networks revisited. *Neuroscientist*, http://dx.doi.org/10.1177/2F1073858416667720.

Bassett, D. S., Meyer-Lindenberg, A., Achard, S., Duke, T., & Bullmore, E. (2006). Adaptive reconfiguration of fractal small-world human brain functional networks. *Proceedings of the National Academy of Sciences*, 103(51), 19518–19523.

Begon, M., Howarth, R. W., & Townsend, C. R. (2014). *Essentials of ecology*. New York: Wiley.

Beisner, B. E., Haydon, D. T., & Cuddington, K. (2003). Alternative stable states in ecology. *Frontiers in Ecology and the Environment*, 1(7), 376–382.

Bellamy Foster, J., B. Clark, and R. York. (2010). *The ecological rift: Capitalism's war on the earth*. New York: Monthly Review Press.

Belsky, A. J. (1986). Does herbivory benefit plants? A review of the evidence. *American Naturalist*, 127(6), 870–892.

Benincà, E., Jöhnk, K. D., Heerkloss, R., & Huisman, J. (2009). Coupled predator–prey oscillations in a chaotic food web. *Ecology Letters*, 12(12), 1367–1378.

Bennett, M., Schatz, M. F., Rockwood, H., & Wiesenfeld, K. (2002). Huygens's clocks. *Proceedings: Mathematics, Physical and Engineering Sciences*, 563–579.

Bernstein, H. (2009). V. I. Lenin and A. V. Chayanov: Looking back, looking forward. *Journal of Peasant Studies*, 36(1), 55–81.

Binder, P. M. (2008). Philosophy of science: Theories of almost everything. *Nature*, 455(7215), 884–885.

Blasius, B., Huppert, A., & Stone, L. (1999). Complex dynamics and phase synchronization in spatially extended ecological systems. *Nature*, 399(6734), 354–359.

Bohlen, P. J., & House, G. (Eds.). (2009). *Sustainable agroecosystem management: integrating ecology, economics, and society*. London: CRC Press.

Byrne, D., & Callaghan, G. (2013). *Complexity theory and the social sciences: The state of the art*. New York: Routledge.

Cafaro, P. (2013). Rachel Carson's environmental ethics. In *Linking ecology and ethics for a changing world* (pp. 163–171). Springer: the Netherlands.

Campbell, H. M., ed. (2009). *The Britannica guide to political science and social movements that changed the modern world*. Rosen, pp. 127–129.

Cannon, S. (1978). *Science in culture*. New York: Dawson and Science History.

Carpenter, S. R., Cole, J. J., Pace, M. L., Batt, R., Brock, W. A., Cline, T., . . . & Smith, L. (2011). Early warnings of regime shifts: A whole-ecosystem experiment. *Science*, 332(6033), 1079–1082.

Carroll, C. R., Vandermeer, J. H., & Rosset, P. M. (1990). *Agroecology*. New York: McGraw-Hill Inc.

Chase, J. M., & Leibold, M. A. (2003). *Ecological niches: Linking classical and contemporary approaches*. Chicago: University of Chicago Press.

Chen, Y. (2013). Loop: An R package for performing decomposition of weighted directed graphs, food web analysis and flexible network plotting. *Ecological Informatics*, 13, 17–21.

Clements, F. E. (1928). *Plant succession and indicators*. New York: H. W. Wilson.

Conway, G. R., & Barbier, E. (1990). *After the green revolution, Sustainable agriculture for development*. London: Earthscan.

Costantino, R. F., Desharnais, R. A., Cushing, J. M., & Dennis, B. (1997). Chaotic dynamics in an insect population. *Science*, 275(5298), 389–391.

Costantino, R. F., Desharnais, R. A., Cushing, J. M., Dennis, B., Henson, S. M., & King, A. A. (2005). Nonlinear stochastic population dynamics: The flour beetle *Tribolium* as an effective tool of discovery. *Advances in Ecological Research*, 37, 101–141.

Crews, T. E., & Gliessman, S. R. (1991). Raised field agriculture in Tlaxcala, Mexico: An ecosystem perspective on maintenance of soil fertility. *American Journal of Alternative Agriculture*, 6(1), 9–16.

Cushing, J. M. (2007). Nonlinearity and stochasticity in population dynamics. In *Mathematics for ecology and environmental sciences* (pp. 125–144). Berlin Heidelberg: Springer.

Darwin, C. (1859). *On the origin of species.* London: John Murray.

Dawson, T., & Fry, R. (1998). Agriculture in nature's image. *Trends in Ecology & Evolution,* 13(2), 50–51.

Delgadillo Polanco, V. M. (2009). Patrimonio urbano y turismo cultural en la Ciudad de México: Las Chinampas de Xochimilco y el Centro Histórico. *Andamios,* 6(12), 69–94.

Demarest, A. (2004). *Ancient Maya: The rise and fall of a rainforest civilization* (Vol. 3). Cambridge: Cambridge University Press.

Desmarais, A. A. (2012). *La Vía Campesina.* Upper Saddle River, NJ: John Wiley & Sons.

De Steven, D. (1982). Seed production and seed mortality in a temperate forest shrub (witch-hazel, *Hamamelis virginiana*). *Journal of Ecology,* 437–443.

Dettelbach, M. (1999). The face of nature: Precise measurement, mapping, and sensibility in the work of Alexander von Humboldt. *Studies in History and Philosophy of Science Part C: Studies in History and Philosophy of Biological and Biomedical Sciences,* 30(4), 473–504.

Egerton, F. N. (2011). History of ecological sciences, part 40: Darwin's evolutionary ecology. *Bulletin of the Ecological Society of America,* 92(4), 351–374.

Ellis, F. (1993). *Peasant economics: Farm households in agrarian development* (Vol. 23). Cambridge: Cambridge University Press.

Elton, C. S. (2001). *Animal ecology.* Chicago: University of Chicago Press.

Ettema, C. H., & Wardle, D. A. (2002). Spatial soil ecology. *Trends in Ecology & Evolution,* 17(4), 177–183.

Fischer, J., Brosi, B., Daily, G. C., Ehrlich, P. R., Goldman, R., Goldstein, J., . . . & Ranganathan, J. (2008). Should agricultural policies encourage land sparing or wildlife-friendly farming? *Frontiers in Ecology and the Environment,* 6(7), 380–385.

Francis, C. A. (1986). *Multiple cropping systems* (No. 631.58 M961). New York: Macmillan.

Francis, C. A., Jordan, N., Porter, P., Breland, T. A., Lieblein, G., Salomonsson, L., . . . Langer, V. (2011). Innovative education in agroecology: Experiential learning for a sustainable agriculture. *Critical Reviews in Plant Sciences,* 30(1–2), 226–237.

Frean, M., & Abraham, E. R. (2001). Rock-scissors-paper and the survival of the weakest. *Proceedings of the Royal Society of London B: Biological Sciences,* 268(1474), 1323–1327.

Friedmann, H. (1980). Household production and the national economy: Concepts for the analysis of agrarian formations. *Journal of Peasant Studies,* 7(2), 158–184.

Garcia-Barrios, L. (2003). Plant–plant interactions in tropical agriculture. *Tropical Agroecosystems,* 11–58.

Glaum, P. R., Ives, A. R., & Andow, D. A. (2012). Contamination and management of resistance evolution to high-dose transgenic insecticidal crops. *Theoretical Ecology,* 5(2), 195–209.

Glaum, P., & Vandermeer, J. (2015). Potential for and consequences of naturalized Bt products: Qualitative dynamics from indirect intransitivities. *Ecological Modelling,* 299, 121–129.

Gliessman, S. R. (1987). Species interactions and community ecology in low external-input agriculture. *American Journal of Alternative Agriculture,* 2(4), 160–165.

Gliessman, S. R. (1990). Agroecology: researching the ecological basis for sustainable agriculture. In *Agroecology* (pp. 3-10). New York: Springer.

Gliessman, S. R. (2006). *Agroecology: the ecology of sustainable food systems.* London: CRC Press.

Gliessman, S. R., Garcia, R. E., & Amador, M. A. (1981). The ecological basis for the application of traditional agricultural technology in the management of tropical agro-ecosystems. *Agro-Ecosystems,* 7(3), 173–185.

Goldberg, D. E., & Landa, K. (1991). Competitive effect and response: Hierarchies and correlated traits in the early stages of competition. *Journal of Ecology,* 1013–1030.

Golubski, A. J., & Abrams, P. A. (2011). Modifying modifiers: What happens when interspecific interactions interact? *Journal of Animal Ecology*, 80(5), 1097–1108.

Golubski, A. J., Westlund, E. E., Vandermeer, J., & Pascual, M. (2016). Ecological networks over the edge: Hypergraph trait-mediated indirect interaction (TMII) structure. *Trends in Ecology & Evolution*, 31(5), 344–354.

Gómez-Pompa, A. (1997). Biodiversity and agriculture: Friends or foes. In *Proceedings of the 1st Sustainable Coffee Congress (September 1996)* (pp. 1–17). Washington, DC.

Gómez Pompa, A., & Venegas, R. (1982). *La chinampa tropical* (No. F/639.9 I5/5).

Goodman, D., DuPuis, E. M., & Goodman, M. K. (2012). *Alternative food networks: Knowledge, practice, and politics*. New York: Routledge.

Gotelli, N. J. (2008). *A primer of ecology*. Sunderland, MA: Sinauer.

Grinnell, J. (1924). Geography and evolution. *Ecology*, 5(3), 225–229.

Hanski, I. (1999). *Metapopulation ecology*. Oxford: Oxford University Press.

Harper, J. L. (1977). *Population biology of plants*. London: Academic Press.

Harwood, R. R. (1990). A history of sustainable agriculture. *Sustainable Agricultural Systems*, 3–19.

Hecht, S. (2010). The new rurality: Globalization, peasants and the paradoxes of landscapes. *Land Use Policy*, 27(2), 161–169.

Heckman, J. (2006). A history of organic farming: Transitions from Sir Albert Howard's War in the Soil to USDA National Organic Program. *Renewable Agriculture and Food Systems*, 21(03), 143–150.

Henson, S. M., Costantino, R. F., Cushing, J. M., Desharnais, R. A., Dennis, B., & King, A. A. (2001). Lattice effects observed in chaotic dynamics of experimental populations. *Science*, 294(5542), 602–605.

Henson, S. M., King, A. A., Costantino, R. F., Cushing, J. M., Dennis, B., & Desharnais, R. A. (2003). Explaining and predicting patterns in stochastic population systems. *Proceedings of the Royal Society of London B: Biological Sciences*, 270(1524), 1549–1553.

HilleRisLambers, R., Rietkerk, M., van den Bosch, F., Prins, H. H., & de Kroon, H. (2001). Vegetation pattern formation in semi-arid grazing systems. *Ecology*, 82(1), 50–61.

Hirota, M., Holmgren, M., Van Nes, E. H., & Scheffer, M. (2011). Global resilience of tropical forest and savanna to critical transitions. *Science*, 334(6053), 232–235.

Holland, J. H. (2012). *Signals and boundaries: Building blocks for complex adaptive systems*. Cambridge, MA: MIT Press.

Holland J.H. (2014). *Complexity: A very short introduction*. Oxford: Oxford University Press.

Holt-Giménez, E. (2009). From food crisis to food sovereignty: The challenge of social movements. *Monthly Review*, 61(3), 142.

Hopper, K. R., & Roush, R. T. (1993). Mate finding, dispersal, number released, and the success of biological control introductions. *Ecological Entomology*, 18(4), 321–331.

Howard, L. E. (1953). *Sir Albert Howard in India*. London: Faber & Faber.

Hrbáček, J., Dvořakova, M., Kořínek, V., & Procházková, L. (1961). Demonstration of the effect of the fish stock on the species composition of zooplankton and the intensity of metabolism of the whole plankton association. *Verhandlungen des Internationalen Verein Limnologie*, 14, 192–195.

Hsieh, H. Y., Liere, H., Soto, E. J., & Perfecto, I. (2012). Cascading trait-mediated interactions induced by ant pheromones. *Ecology and Evolution*, 2(9), 2181–2191.

Huang, Y., & Diekmann, O. (2003). Interspecific influence on mobility and Turing instability. *Bulletin of Mathematical Biology*, 65(1), 143–156.

Huffaker, C. B. (1958). Experimental studies on predation: Dispersion factors and predator-prey oscillations. *Hilgardia*, 27, 343–383.

Huke, J. P. (2006). *Embedding nonlinear dynamical systems: A guide to Takens' theorem*. Crown Copyright, 1993, Manchester Institute for Mathematical Sciences.

Humphries, M. D., & Gurney, K. (2008). Network "small-world-ness": A quantitative method for determining canonical network equivalence. *PloS One*, 3(4), e0002051.

Hunter, M. (2016). *The phytochemical landscape*. Princeton: Princeton University Press.

Hutchinson, G. E. (1957). Cold Spring Harbor Symposium on Quantitative Biology. Concluding Remarks, 22, 415–427.

Huxley, T. H. (1892). *Essays upon some controverted questions*. New York: Macmillan.

Jackson, D., Allen, D., Perfecto, I., & Vandermeer, J. (2014). Self-organization of background habitat determines the nature of population spatial structure. *Oikos*, 123(6), 751–761.

Jackson, D., Skillman, J., & Vandermeer, J. (2012). Indirect biological control of the coffee leaf rust, *Hemileia vastatrix*, by the entomogenous fungus *Lecanicillium lecanii* in a complex coffee agroecosystem. *Biological Control*, 61(1), 89–97.

Jackson, D., Vandermeer, J., & Perfecto, I. (2009). Spatial and temporal dynamics of a fungal pathogen promote pattern formation in a tropical agroecosystem. *Open Ecology Journal*, 2, 62–73.

Jackson, D. L. (Ed.). (2002). *The farm as natural habitat: Reconnecting food systems with ecosystems*. Washington, DC: Island Press.

Jackson, D. W., Zemenick, K., & Huerta, G. (2012a). Occurrence in the soil and dispersal of *Lecanicillium lecanii*, a fungal pathogen of the green coffee scale (*Coccus viridis*) and coffee rust (*Hemileia vastatrix*). *Tropical and Subtropical Agroecosystems*, 15(2).

Jackson, W. (1980). *New roots for agriculture*. Lincoln: University of Nebraska Press.

Jackson, W. (2002). Natural systems agriculture: A truly radical alternative. *Agriculture, Ecosystems & Environment*, 88(2), 111–117.

Jillson, D. (1980). Insect populations respond to fluctuating environments. *Nature*, 288, 699–700.

Johnson, N. (2009). *Simply complexity: a clear guide to complexity theory*. London: Oneworld.

Justus, J. (2005). Qualitative scientific modeling and loop analysis. *Philosophy of Science*, 72(5), 1272–1286.

Kareiva, P. (1982). Experimental and mathematical analyses of herbivore movement: quantifying the influence of plant spacing and quality on foraging discrimination. *Ecological Monographs*, 52(3), 261–282.

Kéfi, S., Rietkerk, M., Alados, C. L., Pueyo, Y., Papanastasis, V. P., ElAich, A., & De Ruiter, P. C. (2007). Spatial vegetation patterns and imminent desertification in Mediterranean arid ecosystems. *Nature*, 449(7159), 213–217.

Kéfi, S., Rietkerk, M., Roy, M., Franc, A., De Ruiter, P. C., & Pascual, M. (2011). Robust scaling in ecosystems and the meltdown of patch size distributions before extinction. *Ecology Letters*, 14(1), 29–35.

King, A. A., Costantino, R. F., Cushing, J. M., Henson, S. M., Desharnais, R. A., & Dennis, B. (2004). Anatomy of a chaotic attractor: Subtle model-predicted patterns revealed in population data. *Proceedings of the National Academy of Sciences*, 101(1), 408–413.

King, A. A., & Schaffer, W. M. (1999). The rainbow bridge: Hamiltonian limits and resonance in predator-prey dynamics. *Journal of Mathematical Biology*, 39(5), 439–469.

Krebs, C. J. (2008). *The experimental analysis of distribution and abundance*. New York: Pearson.

Laland, K. N., Uller, T., Feldman, M. W., Sterelny, K., Müller, G. B., Moczek, A., . . . & Odling-Smee, J. (2015, August). The extended evolutionary synthesis: Its structure, assumptions and predictions. In *Proceedings of the Royal Society of London B*, 282(1813).

Lavigne, D. (2003). Marine mammals and fisheries: The role of science in the culling debate. In Gales, N., Hindell, M., & Kirkwood, R. (eds.), *Marine mammals: Fisheries, tourism and management issues.* Collingwood, Victoria, Australia: CSIRO.

Lear, L. (1998). *Rachel Carson: Witness for Nature.* New York: Macmillan.

Leibold, M. A. (1995). The niche concept revisited: Mechanistic models and community context. *Ecology,* 76(5), 1371–1382.

Leopold, A. (1949). *A Sand County almanac.* Oxford: Oxford University Press.

Leterme, S. C., Seuront, L., & Edwards, M. (2006). Differential contribution of diatoms and dinoflagellates to phytoplankton biomass in the NE Atlantic Ocean and the North Sea. *Marine Ecology Progress Series,* 312, 57–65.

Levin, S. A., Powell, T. M., & Steele, J. H. (Eds.). (2012). *Patch dynamics* (Vol. 96). New York: Springer Science & Business Media.

Levins, R. (1968). *Evolution in changing environments: Some theoretical explorations* (No. 2). Princeton: Princeton University Press.

Levins, R. (1969a). The effect of random variations of different types on population growth. *Proceedings of the National Academy of Sciences,* 62(4), 1061–1065.

Levins, R. (1969b). Some demographic and genetic consequences of envireonmental heterogeneity for biological control. *Bulletin of the Entomological Society of America,* 15, 237–240.

Levins, R. (1974). Discussion paper: The qualitative analysis of partially specified systems. *Annals of the New York Academy of Sciences,* 231(1), 123–138.

Levins, R. (1979). Coexistence in a variable environment. *American Naturalist,* 114(6), 765–783.

Levins, R., & Heatwole, H. (1963). On the distribution of organisms on islands. *Caribbean Journal of Science,* 3, 173–177.

Levins, R., & Lewontin, R. (1985). *The dialectical biologist.* Cambridge, MA: Harvard University Press.

Lewis, W. J., Van Lenteren, J. C., Phatak, S. C., & Tumlinson, J. H. (1997). A total system approach to sustainable pest management. *Proceedings of the National Academy of Sciences,* 94(23), 12243–12248.

Lewontin, R. C., & Cohen, D. (1969). On population growth in a randomly varying environment. *Proceedings of the National Academy of Sciences,* 62(4), 1056–1060.

Liere, H., Jackson, D., & Vandermeer, J. (2012). Ecological complexity in a coffee agroecosystem: Spatial heterogeneity, population persistence and biological control. *PloS One,* 7(9), e45508.

Liere, H., & Larsen, A. (2010). Cascading trait-mediation: Disruption of a trait-mediated mutualism by parasite-induced behavioral modification. *Oikos,* 119(9), 1394–1400.

Liere, H., Perfecto, I., & Vandermeer, J. (2014). Stage-dependent responses to emergent habitat heterogeneity: Consequences for a predatory insect population in a coffee agroecosystem. *Ecology and Evolution,* 4(16), 3201–3209.

Lindeman, R. L. (1942). The trophic-dynamic aspect of ecology. *Ecology,* 23(4), 399–417.

Loewenstein, D. (2001). *Representing revolution in Milton and his contemporaries: religion, politics, and polemics in radical puritanism.* Cambridge: Cambridge University Press.

MacArthur, R., & Levins, R. (1967). The limiting similarity, convergence, and divergence of coexisting species. *American Naturalist,* 101(921), 377–385.

MacArthur, R. H., & Wilson, E. O. (1967). *Theory of island biogeography (MPB-1)* (Vol. 1). Princeton University Press.

May, R. M. (1971). Stability in multispecies community models. *Mathematical Biosciences,* 12(1–2), 59–79.

May, R. M. (1973a). Qualitative stability in model ecosystems. *Ecology,* 54(3), 638–641.

May, R. M. (1973b). *Stability and complexity in model ecosystems* (Vol. 6). Princeton: Princeton University Press.

May, R. M. (1974). On the theory of niche overlap. *Theoretical Population Biology,* 5(3), 297–332.

McCook, S. (2006). Global rust belt: *Hemileia vastatrix* and the ecological integration of world coffee production since 1850. *Journal of Global History,* 1(2), 177–195.

McCook, S., & Vandermeer, J. (2015). The big rust and the red queen: Long-term perspectives on coffee rust research. *Phytopathology,* 105(9), 1164–1173.

McGuire, R. H. (1982). *Hohokam and Patayan: Prehistory of southwestern Arizona.* New York: Academic Press.

McKey, D., Rostain, S., Iriarte, J., Glaser, B., Birk, J. J., Holst, I., & Renard, D. (2010). Pre-Columbian agricultural landscapes, ecosystem engineers, and self-organized patchiness in Amazonia. *Proceedings of the National Academy of Sciences,* 107(17), 7823–7828.

McLoone, M. (1998). *George Washington Carver.* Mankato, MN: Capstone.

Miller, J. H., & Page, S. E. (2007). *Complex adaptive systems: An introduction to computational models of social life.* Princeton: Princeton University Press.

Mitchell, M. (2009). *Complexity: A guided tour.* Oxford University Press.

Moguel, P., & Toledo, V. M. (1999). Biodiversity conservation in traditional coffee systems of Mexico. *Conservation Biology,* 13(1), 11–21.

Morales, H., & Perfecto, I. (2000). Traditional knowledge and pest management in the Guatemalan highlands. *Agriculture and Human Values,* 17(1), 49–63.

Murray, J. D. (1990). Discussion: Turing's theory of morphogenesis: Its influence on modelling biological pattern and form. *Bulletin of Mathematical Biology,* 52(1–2), 119–152.

Newman, M. E., & Watts, D. J. (1999). Renormalization group analysis of the small-world network model. *Physics Letters A,* 263(4), 341–346.

Neyman, J., Park, T., & Scott, E. L. (1956). Struggle for existence: The Tribolium model: Biological and statistical aspects. In *Proceedings of the Third Berkeley Symposium on Mathematical Statistics and Probability* (Vol. 4, No. 4). Berkeley: University of California Press.

Odling-Smee, F. J., Laland, K. N., & Feldman, M. W. (2003). *Niche construction: The neglected process in evolution* (No. 37). Princeton: Princeton University Press.

Oelhaf, R. C. (1978). *Organic agriculture: Economic and ecological comparisons with conventional methods.* Upper Saddle River, NJ: John Wiley and Sons.

Ohgushi, T., Schmitz, O., & Holt, R. (Eds.). (2012). *Trait-mediated indirect interactions.* Cambridge: Cambridge University Press.

Olea, L., & San Miguel-Ayanz, A. (2006). The Spanish dehesa: A traditional Mediterranean silvopastoral system linking production and nature conservation. *Grassland Science in Europe,* 11, 3–13.

Ong, T. W. Y., & Vandermeer, J. H. (2015). Coupling unstable agents in biological control. *Nature Communications,* 6.

Padoch, C., & Pinedo-Vasquez, M. (2010). Saving slash-and-burn to save biodiversity. *Biotropica,* 42(5), 550–552.

Park, T. (1954). Experimental studies of interspecies competition II: Temperature, humidity, and competition in two species of *Tribolium. Physiological Zoology,* 27(3), 177–238.

Pascual, M., & Guichard, F. (2005). Criticality and disturbance in spatial ecological systems. *Trends in Ecology & Evolution,* 20(2), 88–95.

Peacor, S. D., & Werner, E. E. (2001). The contribution of trait-mediated indirect effects to the net effects of a predator. *Proceedings of the National Academy of Sciences,* 98(7), 3904–3908.

Peng, G. S., Tan, S. Y., Wu, J., & Holme, P. (2016). Trade-offs between robustness and small-world effect in complex networks. *Scientific Reports,* 6.

Perfecto, I., & Vandermeer, J. (2015a). *Coffee agroecology: A new approach to understanding agricultural biodiversity, ecosystem services and sustainable development.* New York: Earthscan, Routledge.

Perfecto, I., & Vandermeer, J. (2015b). Structural constraints on novel ecosystems in agriculture: The rapid emergence of stereotypic modules. *Perspectives in Plant Ecology, Evolution and Systematics,* 17(6), 522–530. http://dx.doi.org/10.1016/j.ppees.2015.09.002.

Perfecto, I., Vandermeer, J., & Philpott, S. M. (2014). Complex ecological interactions in the coffee agroecosystem. *Annual Review of Ecology, Evolution, and Systematics,* 45, 137–158.

Piper, J. K. (1999). Natural systems agriculture. *Biodiversity in Agroecosystems,* 167–195.

Polis, G. A. (1991). Complex trophic interactions in deserts: An empirical critique of food-web theory. *American Naturalist,* 138, 123–155.

Power, A. G. (1989). Influence of plant spacing and nitrogen fertilization in maize on Dalbulus maidis (Homoptera: Cicadellidae), vector of corn stunt. *Environmental Entomology,* 18(3), 494–498.

Pretty, J. N. (1995). *Regenerating agriculture: Policies and practice for sustainability and self-reliance.* London: Joseph Henry Press.

Pretty, J., & Buck, L. (2002). Social capital and social learning in the process of natural resource management. In *Natural resources management in African agriculture* (pp. 23–33). Walingford, UK: CAB.

Prigogine, I., & Stengers, I. (1997). *The end of certainty.* New York: Simon and Schuster.

Ramert, B., Lennartsson, M., & Davies, G. (2002). The use of mixed species cropping to manage pests and diseases–theory and practice. In *Proceedings of the UK Organic Research 2002 Conference* (pp. 207–210). Organic Centre Wales, Institute of Rural Studies, University of Wales, Aberystwyth.

Ranzani, L., & Aumentado, J. (2015). Graph-based analysis of nonreciprocity in coupled-mode systems. *New Journal of Physics,* 17(2), 023024.

Renard, D., Iriarte, J., Birk, J. J., Rostain, S., Glaser, B., & McKey, D. (2012). Ecological engineers ahead of their time: The functioning of pre-Columbian raised-field agriculture and its potential contributions to sustainability today. *Ecological Engineering,* 45, 30–44.

Rickerl, D., & Francis, C. (2004a). Agroecosystems analysis. In Rickerl, D. & Francis, C. (eds.), *American Society of Agronomy.*

Rickerl, D., & Francis, C. (2004b). Multi-dimensional thinking: A prerequisite to agroecology. In Rickerl, D. & Francis, C. (eds.), *American Society of Agronomy* (pp. 1–18).

Ricklefs, R., & Relyea, R. (2013). *Ecology: The economy of nature.* New York: W. H. Freeman.

Rietkerk, M., Dekker, S. C., de Ruiter, P. C., & van de Koppel, J. (2004). Self-organized patchiness and catastrophic shifts in ecosystems. *Science,* 305(5692), 1926–1929.

Risch, S. J. (1981). Insect herbivore abundance in tropical monocultures and polycultures: an experimental test of two hypotheses. *Ecology,* 62(5), 1325–1340.

Risch, S. J., Andow, D., & Altieri, M. A. (1983). Agroecosystem diversity and pest control: data, tentative conclusions, and new research directions. *Environmental Entomology,* 12(3), 625–629.

Rival, L. M. (2005). The growth of family trees: Understanding Huaorani perceptions of the forest. In Surallés, A. & García Hierro, P. (eds.), *The land within: Indigenous territory and perception of the environment.* Skive, Denmark: Centraltrykkeriet Skive A/S.

Rival, L. M. (2012). *Trekking through history: The Huaorani of Amazonian Ecuador.* New York: Columbia University Press.

Robertson, G. P., & Freckman, D. W. (1995). The spatial distribution of nematode trophic groups across a cultivated ecosystem. *Ecology,* 76(5), 1425–1432.

Root, R. B. (1973). Organization of a plant-arthropod association in simple and diverse habitats: the fauna of collards (*Brassica oleracea*). *Ecological Monographs*, 43(1), 95–124.

Root, R. B., & Kareiva, P. M. (1984). The search for resources by cabbage butterflies (*Pieris rapae*): ecological consequences and adaptive significance of Markovian movements in a patchy environment. *Ecology*, 65(1), 147–165.

Rosado-May, F. J. (2015). The intercultural origin of agroecology: Contributions from Mexico. In *Agroecology: A transdisciplinary, participatory and action-oriented approach* (pp. 123–138). CRC Press.

Rosset, P. (2013). Re-thinking agrarian reform, land and territory in La Via Campesina. *Journal of Peasant Studies*, 40(4), 721–775.

Rosset, P. M., & Altieri, M. A. (1997). Agroecology versus input substitution: A fundamental contradiction of sustainable agriculture. *Society & Natural Resources*, 10(3), 283–295.

Rosset, P. M., & Martinez-Torres, M. E. (2013). La Via Campesina and agroecology. *La Via Campesina's open book: Celebrating*, 20, 1–22.

Rosset, P. M., & Martínez-Torres, M. E. (2014). Food sovereignty and agroecology in the convergence of rural social movements. In *Alternative agrifood movements: Patterns of convergence and divergence* (pp. 137–157). Emerald Group.

Russell, E. P. (1989). Enemies hypothesis: a review of the effect of vegetational diversity on predatory insects and parasitoids. *Environmental Entomology*, 18(4), 590–599.

Sahimi, M. (1994). *Applications of percolation theory*. London: CRC Press.

Sakai, K. (2001). *Nonlinear dynamics and chaos in agricultural systems*. London: Elsevier.

Sakai, K., Managi, S., Vitanov, N. K., & Demura, K. (2007). Transition of chaotic motion to a limit cycle by intervention of economic policy: An empirical analysis in agriculture. *Nonlinear Dynamics, Psychology and Life Sciences*, 11, 253–265.

Sakai, K., Noguchi, Y., & Asada, S. I. (2008). Detecting chaos in a citrus orchard: Reconstruction of nonlinear dynamics from very short ecological time series. *Chaos, Solitons & Fractals*, 38(5), 1274–1282.

Sanz-Arigita, E. J., Schoonheim, M. M., Damoiseaux, J. S., Rombouts, S. A., Maris, E., Barkhof, F., . . . & Stam, C. J. (2010). Loss of "small-world" networks in Alzheimer's disease: Graph analysis of FMRI resting-state functional connectivity. *PloS One*, 5(11), e13788.

Scheffer, M. (2009). *Critical transitions in nature and society*. Princeton: Princeton University Press.

Scheffer, M., Carpenter, S. R., Lenton, T. M., Bascompte, J., Brock, W., Dakos, V., . . . & Vandermeer, J. (2012). Anticipating critical transitions. *Science*, 338(6105), 344–348.

Schindler, D. E., Carpenter, S. R., Cottingham, K. L., He, X., Hodgson, J. R., Kitchell, J. F., & Soranno, P. A. (1996). Food web structure and littoral zone coupling to pelagic trophic cascades. In *Food webs* (pp. 96–105). New York: Springer.

Schmidt, M. W., Torn, M. S., Abiven, S., Dittmar, T., Guggenberger, G., Janssens, I. A., . . . Nannipieri, P. (2012). Persistence of soil organic matter as an ecosystem property. *Nature*, 478(7367), 49–56.

Simberloff, D. (1980). A succession of paradigms in ecology: essentialism to materialism and probabilism. *Synthese*, 43, 3–9.

Si-ming, L.M.W. (2011). Survey and practice of agro-cultural heritage protection in Jiangsu [J]. *Agricultural History of China*, 1, 19.

Soemarwoto, O., & Conway, G. R. (1992). The Javanese homegarden. *Journal for Farming Systems Research-Extension*, 2(3), 95–118.

St. Barbe Baker, R. (1948). *Green glory: The story of the forests of the world*.

Stanhill, G. (1990). The comparative productivity of organic agriculture. *Agriculture, Ecosystems & Environment*, 30(1), 1–26.

Stauffer, D., & Aharony, A. (1994). *Introduction to percolation theory*. London: CRC Press.

Staver, A. C., Archibald, S., & Levin, S. (2011a). Tree cover in sub-Saharan Africa: Rainfall and fire constrain forest and savanna as alternative stable states. *Ecology*, 92(5), 1063–1072.

Staver, A. C., Archibald, S., & Levin, S. A. (2011b). The global extent and determinants of savanna and forest as alternative biome states. *Science*, 334(6053), 230–232.

Stige, L. C., Chan, K. S., Zhang, Z., Frank, D., & Stenseth, N. C. (2007). Thousand-year-long Chinese time series reveals climatic forcing of decadal locust dynamics. *Proceedings of the National Academy of Sciences*, 104(41), 16188–16193.

Strogatz, S. (2003). *Sync: The emerging science of spontaneous order*. New York: Hyperion.

Strong, D. R. (1986). Density-vague population change. *Trends in Ecology & Evolution*, 1(2), 39–42.

Swift, M. J., Izac, A. M., & van Noordwijk, M. (2004). Biodiversity and ecosystem services in agricultural landscapes: Are we asking the right questions? *Agriculture, Ecosystems & Environment*, 104(1), 113–134.

Swift, M. J., Vandermeer, J., Ramakrishnan, P. S., Anderson, J. M., Ong, C. K., & Hawkins, B. A. (1996). Biodiversity and agroecosystem function. *Scope-Scientific Committee on Problems of the Environment International Council of Scientific Unions*, 55, 261–298.

Tahvanainen, J. O., & Root, R. B. (1972). The influence of vegetational diversity on the population ecology of a specialized herbivore, Phyllotreta cruciferae (Coleoptera: Chrysomelidae). *Oecologia*, 10(4), 321–346.

Tainter, J. (1990). *The collapse of complex societies*. Cambridge: Cambridge University Press.

Tainter, J. A. (2006). Archaeology of overshoot and collapse. *Annual Review of Anthropology*, 35, 59–74.

Takens, F. (1981). Detecting strange attractors in turbulence. In *Dynamical systems and turbulence, Warwick 1980* (pp. 366–381). Berlin Heidelberg: Springer.

Ter Hark, M. (2009). Popper's theory of the searchlight: A historical assessment of its significance. In *Rethinking Popper* (pp. 175–184). Springer Netherlands.

Thom, R. (1989). *Structural stability and morphogenesis: An outline of a general theory of models*. Reading, MA Addison-Wesley.

Tilley, H. (2011). *Africa as a living laboratory: Empire, development, and the problem of scientific knowledge, 1870–1950*. Chicago: University of Chicago Press.

Tilman, D., & Kareiva, P. M. (1997). *Spatial ecology: The role of space in population dynamics and interspecific interactions* (Vol. 30). Princeton: Princeton University Press.

Travers, J., & Milgram, S. (1967). The small world problem. *Psychology Today*, 1, 61–67.

Tscharntke, T., Klein, A. M., Kruess, A., Steffan-Dewenter, I., & Thies, C. (2005). Landscape perspectives on agricultural intensification and biodiversity: Ecosystem service management. *Ecology Letters*, 8(8), 857–874.

Vandermeer, J. (1989). *The ecology of intercropping*. Cambridge: Cambridge University Press.

Vandermeer, J. (1990). Notes on agroecosystem complexity: Chaotic price and production trajectories deducible from simple one-dimensional maps. *Biological Agriculture & Horticulture*, 6(4), 293–304.

Vandermeer, J. (1993). Loose coupling of predator-prey cycles: Entrainment, chaos, and intermittency in the classic MacArthur consumer-resource equations. *American Naturalist*, 141(5), 687–716.

Vandermeer, J. (1995). The ecological basis of alternative agriculture. *Annual Review of Ecology and Systematics*, 201–224.

Vandermeer, J. (2004a). Coupled oscillations in food webs: Balancing competition and mutualism in simple ecological models. *American Naturalist*, 163(6), 857–867.

Vandermeer, J. (2004b). The importance of a constructivist view. *Science*, 303(5657), 472–474.

Vandermeer, J. (2006). Oscillating populations and biodiversity maintenance. *Bioscience*, 56(12), 967–975.

Vandermeer, J., I. Perfecto, and S. M. Philpott. (2008). Clusters of ant colonies and robust criticality in a tropical agroecosystem. *Nature*, 451, 457–459.

Vandermeer, J. (2015). Some complications of the elementary forms of competition in a source/sink and metacommunity context: The role of intransitive loops. *arXiv preprint arXiv:1502.05225*.

Vandermeer, J., & Jackson, D. (2015). Spatial pattern and power function deviation in a cellular automata model of an ant population. *arXiv preprint arXiv:1512.08660*.

Vandermeer, J., & Jackson, D. (2018). Stabilizing coupled intransitive loops: Self-organized spatial structure and disjoint time frames. *Journal of Ecology*.

Vandermeer, J., Jackson, D., & Perfecto, I. (2014). Qualitative dynamics of the coffee rust epidemic: Educating intuition with theoretical ecology. *BioScience, Bit034*.

Vandermeer, J., & Kaufmann, A. (1998). Models of coupled population oscillators using 1-D maps. *Journal of Mathematical Biology*, 37(2), 178–202.

Vandermeer, J., & Perfecto, I. (1995). *Breakfast of biodiversity: The truth about rain forest destruction*. Institute for Food and Development Policy.

Vandermeer, J., & Perfecto, I. (2012). Syndromes of production in agriculture: Prospects for social-ecological regime change. *Ecology and Society*, 17(4), 39.

Vandermeer, J., Perfecto, I., & Philpott, S. (2010). Ecological complexity and pest control in organic coffee production: Uncovering an autonomous ecosystem service. *BioScience*, 60(7), 527–537.

Vandermeer, J., & Rohani, P. (2014). The interaction of regional and local in the dynamics of the coffee rust disease. *arXiv preprint arXiv:1407.8247*.

Vandermeer, J., Rohani, P., & Perfecto, I. (2015). Local dynamics of the coffee rust disease and the potential effect of shade. *arXiv preprint arXiv:1510.05849*.

Vandermeer, J., & Yitbarek, S. (2012). Self-organized spatial pattern determines biodiversity in spatial competition. *Journal of Theoretical Biology*, 300, 48–56.

Vandermeer, J., & Yodzis, P. (1999). Basin boundary collision as a model of discontinuous change in ecosystems. *Ecology*, 80(6), 1817–1827.

Vandermeer, J. H. (1969). The competitive structure of communities: An experimental approach with protozoa. *Ecology*, 50(3), 362–371.

Vandermeer, J. H. (1992). *The ecology of intercropping*. Cambridge: Cambridge University Press.

Vandermeer, J. H. (2011). *The ecology of agroecosystems*. Burlington, MA: Jones & Bartlett Learning.

Vandermeer, J. H., & Goldberg, D. E. (2013). *Population ecology: First principles*. Princeton: Princeton University Press.

Van der Ploeg, J. D. (2009). *The new peasantries: Struggles for autonomy and sustainability in an era of empire and globalization*. New York: Routledge.

Van der Ploeg, J. D. (2013). *Peasants and the art of farming: A Chayanovian manifesto (No. 2)*. Winnipeg: Fernwood.

Vasey, D. E. (2002). *An ecological history of agriculture 10,000 BC to AD 10,000*. West Lafayette, IN: Purdue University Press.

Walker, R.E.D., Pastor, J., & Dewey, B. W. (2010). Litter quantity and nitrogen immobilization cause oscillations in productivity of wild rice (*Zizania palustris* L.) in Northern Minnesota. *Ecosystems*, 13(4), 485–498.

Wallace, D. F. (2010). *Everything and more: A compact history of infinity*. New York: W. W. Norton.

Wang, X. F., & Chen, G. (2003). Complex networks: Small-world, scale-free and beyond. *IEEE Circuits and Systems Magazine*, 3(1), 6–20.

Watts, D. J., & Strogatz, S. H. (1998). Collective dynamics of "small-world" networks. *Nature*, 393(6684), 440–442.

Werner, E. E., & Peacor, S. D. (2003). A review of trait-mediated indirect interactions in ecological communities. *Ecology*, 84(5), 1083–1100.

Whittaker, R. H., Levin, S. A., & Root, R. B. (1973). Niche, habitat, and ecotope. *American Naturalist*, 107(955), 321–338.

Wilken, G. C. (1990). *Good farmers: Traditional agricultural resource management in Mexico and Central America.* Berkeley: University of California Press.

Wilkinson, T. J. (1997). Environmental fluctuations, agricultural production and collapse: A view from bronze age upper Mesopotamia. In *Third millennium BC climate change and old world collapse* (pp. 67–106). Berlin Heidelberg: Springer.

Winfree, A. (1980). *The geometry of biological time.* New York: Springer-Verlag.

Wittman, H. (2009). Reworking the metabolic rift: La Vía Campesina, agrarian citizenship, and food sovereignty. *Journal of Peasant Studies*, 36(4), 805–826.

Wootton, J. T. (1993). Indirect effects and habitat use in an intertidal community: Interaction chains and interaction modifications. *American Naturalist*, 141(1), 71–89.

Wright, A. L., & Wolford, W. (2003). *To inherit the earth: The landless movement and the struggle for a new Brazil.* Oakland, CA: Food First Books.

Ye, X., Sakai, K., Manago, M., Asada, S. I., & Sasao, A. (2007). Prediction of citrus yield from airborne hyperspectral imagery. *Precision Agriculture*, 8(3), 111–125.

Zelenev, V. V., Van Bruggen, A.H.C., & Semenov, A. M. (2000). "Bacwave," a spatial temporal model for traveling waves of bacterial populations in response to a moving carbon source in soil. *Microbial Ecology*, 40(3), 260–272.

Zelenev, V. V., Van Bruggen, A.H.C., & Semenov, A. M. (2005). Short-term wavelike dynamics of bacterial populations in response to nutrient input from fresh plant residues. *Microbial Ecology*, 49(1), 83–93.

Index

Page numbers in italics indicate figures.

abstraction 12
Acromyrmex octospinosus ants 85
agricultural syndromes as hysteretic
    phenomena with tipping points 214–217
Agricultural Testament, An 9
agroecology 6; background in formal
    science 224–227; early revolutionaries
    in 7–12; four pillars of 224–230;
    generalization in 53; Indian farming
    practices in 10; interspecific competition
    and 55–59; messages from the natural
    world for 229–230; political resistance
    in 228–229; scientific basis of 221–233;
    solving farmer's problems in the age of
    ecological complexity 231–233;
    thought-intensive technology 233;
    traditional knowledge in 227–228;
    "whole" of 230–231; *see also* ecological
    complexity; ecology
Allee effect 122–123
amino acids 25
Anker, Peder 226
attractors, chaotic 99–103; generalized
    structures of 106–110; reconstruction
    110–113
Azteca ants 72–74, 78, 87, 117; percolation
    points and power functions 81–83;
    trait-mediated indirect interactions
    188–190

bacteria, nitrifying 27
Bak, Per 76
Baker, Richard St. Barbe 229–230
"balance of nature" 42–43
Balfour, Lady Eve 7–8, 10
basin boundary collisions 218–219
Between Land and Sea 8, 10

biodiversity, loss of 201–205
Bohr, Neils 116
boundary collisions 218–219
Boyle, Robert 222
bucket observations of nature 13

cacao plants 60–61, 86
Cantor, George 102
Cantor sets 99–103
Carson, Rachel 8–10, 15
Carver, George Washington 7–10
chaos 113; attractors, transients and Cantor
    sets 99–103; basin boundary collisions
    218–219; from cooked carrots to chaos
    to attractor reconstruction and 110–113;
    ecological, in the real world 103–104,
    105; farmer math and the intuition
    of 95–99; generalized structure of chaotic
    attractors 106–110; interrelationship
    between stochasticity and 127–130;
    introduction to 89–93; intuition of
    importance of 93–95; oscillatory
    structure in 139–140
Chayanov system 215–216
chemical catalyzing microorganisms 27
chlorophyll 20
Chomsky, Noam 226
Clements, Frederic 11, 38–39, 226
clusters, population 62–63
Coffee Agroecology 70, 180
coffee rust disease 208–214
colloids 25–28
community matrix 163–165
complexity science and ecology 222–224
complex systems *see* ecological complexity
constructivism 38–39
consumer/resource oscillators 136–139

Copernicus 225
couple oscillators *see* oscillators, coupled
criticality and power functions 75–80
critical transitions 219–220; agricultural
    syndromes as hysteretic phenomena
    with tipping points 214–217; basin
    boundary collisions 218–219; coffee
    rust disease 208–214; hysteresis on
    global scale 205–208; inevitability
    of surprise and 200; loss of
    biodiversity 201–205

Darwin, Charles 1, 8, 12, 179, 222; on
    ecological niche 34–35, 37; historical
    view of 225; on trait-mediated
    non-linearities 186
Darwin's Sacred Cause 225
decomposers/decomposition 25–28;
    as oscillatory process 149–152
dehesa system 162–163
denitrifying bacteria 27
density-mediated indirect
    interactions 179–181
Desmond, Adrian 225
deterministic versus stochastic
    science 115–116
Dialectical Biologist, The 6, 12–13
Diggers 8–10, 22, 229
directed graphs, qualitative structure
    of 172–176

earthworms 85
ecological complexity 15–16; idea of 2–7;
    solving farmer's problems in age
    of 231–233; *see also* agroecology
ecological niche 34–39
ecology: basic concepts 19–59; bringing
    materialism into 10–12; complexity
    science and 222–224; ecological niche
    in 34–39; equilibrium, resilience,
    persistence concepts 42–47; feedback in
    dynamic systems 50–59; generalization
    in 12–14; how plants get energy 19–23;
    new enthusiasm about 1; origins of 1–2;
    plants, soil and water 32–34;
    plants getting nutrients from soil 24–27;
    transformation of nutrients in soil 27–32;
    trophic dynamics 47–49, *50*, 51–52;
    see also agroecology
Ecology 11
Einstein, Albert 116
Elton, Charles 35–37, 61
energy: basic trophic dynamics and 47–50;
    obtained by plants 19–23

Engels, Friedrich 8, 24
equilibrium 42–47
equilibrium theory of island
    biogeography 157

farmer math and intuition of chaos 95–99
*Farmers of Forty Centuries: Organic Farming in
    China, Korea and Japan* 227
feedback in dynamic systems 50–59
Feynman, Richard 117
food webs 166–172
Foster, John Bellamy 11
fundamental niche 61–62; skeleton
    of 84–85

Galileo Galilei 223, 225
Gause, Georgy 36, 140
generalizations 12–14, 53
generalized structure of chaotic
    attractors 106–110
Grinnell, Joseph 35, 61

Haeckel, Ernst 1
Harper, John 38
Haughley Experiment 7
Hernandez Xolocotzi, Efrain 227–228
Hogben, Lancelot 11
Howard, Albert 7–10, 95–96, 227
How Nature Works 76
Hrbácek, J. 178
Huaorani people 201
Huffaker, C. B. 67–68, 86
Humboldt, Alexander von 95–97
humus 28
Hunter, Mark 47
Hutchinson, G. E. 35–37, 39, 61
Huxley, Thomas 187
Huygens, Christian 135–136, *136*,
    138, 143
hypernetworks 191–196
hysteresis 205–208; agricultural syndromes
    as hysteretic phenomena with tipping
    points 214–217

idealism 12
Imperial Ecology 226
Indian farming practices 10
induction versus deduction 12
industrial agriculture 1, 6
inevitability of surprise 200
interspecific competition 55–59
intuition 14; of chaos and farmer math
    95–99; of importance of chaos in simple
    system 93–95; theory and 14–15

Janzen/Connell mechanism 72

King, Franklin Hiram 227

La Agricultura Precolombina en Chile y los Paises Vecinos 227
Lancaster, Ray 11
Land Institute 230
Laplace, Pierre-Simon 223–224
La Via Campesina 9, 229
Lavigne, David 2, *3*
Levins, Richard 6, 11–13, 15, 144, 191; on community matrix 163–165; on constructivism 38–39; on metapopulations 64, 67; on thought-intensive technology 233
Levy, Hyman 11
Lewontin, R. 6, 12–13, 15; on constructivism 38–39
limiting similarity 143–148, *149*
Lindeman, Raymond 47, 166, 179
Living Soil, The 7
loop analysis 172–176
Lorenz, Ed 92–93
loss of biodiversity 201–205
Lotka-Volterra approach 160

Macarthur, R. 144
Marx, Karl 24
Marxism 11, 24
materialism in ecology 10–13
mathematical representation of theory 14
May, R. M. 191
mean field model 67
metaphor 12–13
metapopulations 64–69
microorganisms: chemical catalyzing 27; metabolic rate of 29; nitrogen-fixing 24; transformation of nutrients in soil 27–32
Milgram, Stanley 168
Miller, J. H. 6, 15
Moore, Robert 225
Morales, Helda 231
Moran effect 152–156
MST (Movimento dos Trabalhadores Rurais Sem Terra) 22, 228–229
multidimensional hypervolume 35
multidimensionality 157–158, 176; promise of food webs and 166–172; qualitative structure of directed graphs in 172–176; systems 162–166; three-dimensional systems 158–162
Murray, Greg 70
mutualism 184–185, 188

National Geographic 227
Nature 28
Newton, Isaac 2, 92, 116, 225
niche: construction 38; ecological 34–39; fundamental 61–62, 84–85; realized 84–86
nitrifying bacteria 27
nitrogen-fixing microorganisms 24, 43–44
non-random spatial patterns 62–64

oscillators, coupled: basic pattern in consumer/resource oscillators and weak coupling 136–139; confronting Gause's principle with oscillations 140–142, *143*; decomposition as oscillatory process and 149–152; discovery of 135; limiting similarity and species packing with 143–148, *149*; oscillatory structure in chaos 139–140; seasonality and the Moran effect 152–156

packing, species 143–148, *149*
Page, S. E. 6, 15
percolation points and power functions 81–84
persistence 42–47
Phillips, John 11
phosphorus solubilizing microorganisms 27
photosynthesis 20–23, 43–44
Phytochemical Landscape, The 47
plants 32–34; basic trophic dynamics and 47–49, *50*; energy obtained by 19–23; getting nutrients from soil 24–27
Platonic Idealism 11–12
Poincaré, Henri 15, 92
political resistance in agroecology 228–229
Popper, Karl 13, 111
population clusters 62–63; criticality and power functions 75–80
population dynamics 39–42; consequences of background exogenous pattern on 64–69; generation of endogenous pattern on 69–74; limiting similarity and species packing with oscillators 143–148, *149*; non-random spatial patterns 62–64; with stochasticity 119–125
power functions: criticality and 75–80; percolation points and 81–84
predator/prey: background exogenous pattern 64–69; oscillations 149–152; seasonality and the Moran effect 152–156; systems and stochasticity 125–127
predictability 12–14
proteins 25

qualitative structure of directed graphs 172–176

rainfall and hysteresis 205–208
realized niche 84–86
resilience 42–47
root hairs 26–27
Rossler attractor 110

Sakai, Kenshi 104
scarecrows 182–183
scientific basis of agroecology 221–233;
    complexity science and ecology 222–224;
    four pillars of agroecology 224–230;
    solving farmer's problems in age
    of ecological complexity 231–233;
    thought-intensive technology 233;
    "whole" of agroecology 230–231
scientific theory *see* theory
Sea Around Us, The 8, 10
seasonality and the Moran effect
    152–156
self-organized critical systems 76–78, 86
Silent Spring 8, 10
similarity, limiting 143–148, *149*
simile 12–13
Smut, Jan Christian 11
soil: -borne fungal pathogens 120;
    plant nutrients obtained from 24–27;
    texture 32–34; transformation of
    nutrients in 27–32; water change
    in 32–34
Soil Association 7–8
solar radiation 20–21
spatial pattern and agricultural connection
    86–87
species packing 143–148, *149*
Starling murmurations 4–5
stochasticity 134; basic population processes
    with 119–125; importance of 116–119;
    interrelationship between chaos and

127–130; introduction to 115–116;
    predator/prey systems and 125–127
symbiosis 24
syndromes of production 214–217

Takens theorem 111–113
"tangled bank" 12
Tansley, Arthur 11, 226
teacup model of oscillatory structure
    139–140
theory 12–13; intuition and 14–15
thought-intensive technology 233
three-dimensional systems 158–162
tipping points 214–217
trait-mediated indirect interactions
    178–179, 198; basic nonlinearity in
    181–187; consequences of 191;
    density-mediated indirect interactions
    and 179–181; effects in the real world
    187–190; hypernetworks 191–196;
    multiple 196, *197*
transients, chaotic 99–103
transitions, critical *see* critical transitions
Traves, Jeffrey 168
Tribolium castaneum 130–134
trophic dynamics 47–49, *50*, 51–52;
    density-mediated indirect interactions
    and 179
True Levelers 8
Turing, Alan 70–72, 74, 86
two-dimensional systems 158–159

Vandermeer, J. 38

water 32–34
ways of knowing 12
weak coupling 136–139
Werner, Earl 187
Wootton, J. T. 187

zero growth isocline 48–49